ONLY ONE EARTH

BBC BOOKS/EARTHSCAN

The people will live on.
The learning and blundering people will live on.
They will be tricked and sold and again sold
And go back to the nourishing earth for rootholds.

CARL SANDBURG, The People, Yes

Published by **BBC Books**
A division of BBC Enterprises Limited
Woodlands, 80 Wood Lane, London W12 0TT
and **Earthscan**
A division of the International Institute for
Environment and Development
3 Endsleigh Street, London WC1H 0DD

First published 1987

ISBN 0 563 20548 2 (hardback)
ISBN 0 563 20549 0 (paperback)

Typeset in 10 pt Monophoto Sabon by Butler & Tanner Limited
Printed and bound in England by Butler & Tanner Limited,
Frome and London

Colour separations by Technik Limited, Berkhamsted
Colour printing by Chorley and Pickersgill Limited, Leeds

CONTENTS

FOREWORD

by GRO HARLEM BRUNDTLAND:
Prime Minister of Norway;
Chairman, World Commission
on Environment and Development

In the 5000 days between now and the end of this century, over a billion people will be added to the planet. During the next century another 2 to 7 billion people could be added before the size of the human family stabilises at somewhere between 8 and 13 billion people. Thus we and our children must plan to squeeze perhaps two new human worlds into only one Earth. We must plan to support them with the same ecosystems from which we today draw our food, fish, energy, wood products, minerals and other materials.

But these systems are already under tremendous pressure. Our World Commission on Environment and Development began its work in October 1984, a sadly appropriate time. Our beginnings coincided with a horrified world's first glimpse into the depths of the human tragedy unfolding in Africa. Perhaps a million people paid with their lives for failures in national and international development efforts. Decades of misdevelopment – overcultivation, overgrazing, soil erosion, deforestation and short-sighted farming policies – all severely damaged Africa's environment and reduced the continent's ability to grow food. But Africa does not suffer the effects of environmental bankruptcy alone. There are many more hungry farmers on deteriorating soil in any given year in Asia than in Africa. Latin America is rapidly turning its forests into deserts – deserts of hunger and despair for the farmers encouraged to cut the trees.

Nor do developing nations have a monopoly on environmental mismanagement. We in the North have not been wise or efficient managers of our own resources. Our agricultural subsidies encourage farmers to overuse fertilisers and pesticides, to overdraw soil and water accounts. Our factories and our wastes pollute the deep groundwater, water we drink but which we cannot reach to clean.

Our Commission has heard evidence from scientists that the damage caused by acid pollution may be far greater than previously realised. These warnings speak of acidification whose clean-up costs are far beyond economic reach, and whose effects stretch beyond forests and lakes into buildings, bridges and national art and architectural treasures. Yet at least 63 per cent, perhaps as much as 92 per cent, of the sulphur emissions acidifying Norway's lakes and soils come from beyond its own borders. Norway's economic security, and that of other 'downwind' nations such as Canada, Sweden and Czechoslovakia, is threatened by activities over which they have no control. Despite national sovereignty, no nation should be free to pollute the common environment and to inflict severe ecological and economic damage on other states.

The problems we face as a world community are planetary, but not insoluble. Our two greatest resources, land and people, can still redeem the promise of development. If we take care of nature, nature will take care of us. But the huge changes sweeping over us and our biospheres demand fundamental changes in our attitudes, our policies and in the way we run our societies.

We need a wider definition of national security, far beyond the narrow confines of military security, to embrace economic and ecological interdependence, and global environmental hazards. Our new definition will require new restrictions and new forms of co-operation, in the interest of all.

This broadened concept of national security will also embrace our relations with the developing world. In the long run, perhaps in the short run too, environmentally-sound development there also benefits the North. The destruction of the planet's environment is making the world a less stable place, politically, economically and militarily.

As the overburdened land can sustain fewer people in Third World nations, desperate and hungry families search for a way out. Some move to cities which are already bursting at the seams. Others remain in the countryside, where they may become involved in sporadic violence between nomads and settled farmers, or in the organised violence of guerrilla movements. Others flee into neighbouring countries, becoming environmental refugees, placing intolerable burdens on their host nations.

In Oslo, London, Washington, Paris or Moscow – we can have no true security until we direct far more of our efforts into ecological stability in the South. Armaments cannot remove the threats to peace. Sustainable development and wise environmental management can.

But, despite their importance, we have traditionally tacked environmental concerns on to the tail-end of our bureaucratic structures in underfunded and understaffed ministries. These concerns must be tackled in the ministries of finance, planning, trade, agriculture, energy, development, foreign affairs, and indeed in the offices of the prime ministers and presidents.

When we use man-made assets, such as equipment and buildings, we write off the use as depreciation. But we forget to evaluate our environment as productive capital, even though we use it as such. When we clear forests, overharvest fisheries, overwork cropland until the soil erodes, and use our skies as free rubbish bins and our rivers as sewers, we can actually show an increase in our national incomes as measured in Gross National Product. But such strategies are short-sighted and such gains short-lived. We must pay eventually, often more heavily than if we had budgeted the costs in the first place. We need new book-keeping systems with columns reckoning the real costs of environmental degradation.

Sustainable development calls for new imperatives for international co-operation. It requires new international financial systems to cope with national debts which both the lender and the debtor know can never be paid. It requires the removal of protectionist measures against commodities such as sugar, upon the sale of which so many Third World nations depend. Environmental damage, high debts and low returns on commodity sales all encourage nations to overdraw further their environmental accounts in futile attempts to balance national accounts.

These changes must reach deep into our societies. Northern citizens responded with generosity to Africa's crisis, but these same citizens are sadly uninformed about the short-sighted development policies, North and South, behind such crises. I believe that society's dominant belief structures are shifting. We are questioning our brash assurance that people can dominate the Earth, that the world is vast and unlimited, that every problem will somehow be solved by others. We are moving towards an era of citizen participation at all levels in the care of the planet.

After two and a half years of scientific investigations and public hearings on five continents, our Commission has ruthlessly documented the risks to humankind's survival. We have looked beyond the effects of a misused environment and sought the political and economic causes. But we have also documented examples of sustainable development.

The publication of our report in April 1987 coincides with the international release of the television series *Only One Earth*, and with the publication of this book.

The following chapters look at development through the eyes of the people making it happen. Living in different environments, guided by different world views, they all face the same basic challenge of realising their ambitions without destroying the resources their children will need to realise theirs. They are the planet's prime resource and the future's best hope. We politicians in capital cities have the duty of supporting their effort, of constructing frameworks for their success, but we know that the necessary changes cannot be dictated from above. They evolve deep in the hearts of people responding to the elemental vision of life as they see it.

Only One Earth: Living for the Future gives me hope that, as usual, the people are leading the politicians. By providing for the future, the heroes and heroines of this book are improving the present. They have gone further: from their ground-level viewpoints in desert camps, villages, farms, shantytowns and suburban neighbourhoods they have already begun to live by the ethic that the Earth is indeed One.

PREFACE

'I mean, things are horrible; they're unendurable; ghastly things happen. But on some level, everything is all right.'
ROBERT STONE, US novelist

This book, despite the bleakness of the first two chapters, is about hope for the future. It marks a major change in the philosophy behind what might be called the development business, which aims to provide poor people in poor countries with sufficient food, better housing, improved health and at least a reasonable chance of realising their ambitions.

Until now, efforts to develop poorer nations, whether made by the nations themselves or by outsiders, have largely failed. It is true that a few Third World nations are rapidly increasing their national earnings, but these are exceptions which must be offset by the numbing statistics which show that in the world today more people than ever before are hungry, ill and poorly housed. Worse still, many efforts at development are destroying the very resources upon which any future development can be based: soil, water, forests, even the air we breathe. Rather as if it were a short-sighted family trying to improve its life-style by squandering its entire savings, the planet itself is overdrawing its environmental capital accounts and going into the red. The famines in Africa in 1984–5 showed that in some regions environmental bankruptcy has been reached.

There are many complex reasons for this, and many savage ironies are exposed in the process of destruction. One of the deepest ironies is that the projects which governments often label 'development' may be part of this prodigal wasting of the planet's resources. The realisation of this irony, brought to light by events in Africa but also by the more recent drought in parts of India and the chronic hunger in much of Latin America, has forced governments to search for a form of development which is ecologically sustainable. It is a form of development based on the premise that the outer surface of the planet and the people living on that outer surface matter more than annual economic indicators. It is a form of development which lives off the interest of the Earth's resources, rather than squandering the capital.

Only One Earth shows the ways in which development is going wrong. But the primary aim is to give examples of how development can be made to work, and to show how people in nine different countries, rich and poor, have been taking charge of their destinies, improving their lives while living within their environmental means.

I

THE DANGERS OF 'DEVELOPMENT'

'We are stuck with this mildly poisoned planet and its smoky air. We are in for hunger and hard work, the highest stage of poverty.'
PAUL THEROUX, novelist and travel writer

The nations of our planet can be divided, very, very roughly, into the haves and the have-nots. Of the 4·8 billion plus people on Earth, 1·2 billion live in more developed nations and 3·6 billion live in less developed nations, the 'more or less developed' status being awarded almost solely on the basis of economic indicators. The more developed nations produce most of the world's goods and services, but the one-quarter of the population who live in those countries also have a higher consumption: they consume three-quarters of the world's goods.

This basic division has been of growing concern to an increasing number of people; especially throughout the second half of this century, which has seen many former colonies gaining independence and joining the ranks of the 'developing nations', or the Third World, or, more recently, simply 'the South'. (This North–South division, which suggests that to live on the North side of the Equator is to live on 'the right side of the tracks', is misleading. The majority of the most populous developing nations lie to the north of the Equator – in fact 85 per cent of the world's inhabitants live north of the Equator. But the use of the term 'North' to describe rich, industrialised countries, and 'South' to describe all other nations, is a handy, benign shorthand which will be used often in this book.)

Much of the anxiety over 'development' is based on moral concern. Many believe that it is simply not fair for so few to have so much of what is available. The distribution of wealth and resources produces statistics which trouble the consciences of thinking people both North and South. For instance, according to British environmentalist Richard North, the average British cat eats twice as much animal protein every day as the average African, and a third more than the average resident of the South. The cost of keeping that cat – about £170 ($255) per year – is more than the average annual income of the 1 billion people who live in the world's fifteen poorest nations. This seems wrong.

Much of the concern is based on practicalities. Countries which are impoverished both in terms of their national treasuries and their citizens' purchasing power are not good customers for the excess of products and services provided by the Northern industrialised nations. Also, poverty is often accompanied by political instability, *coups*, civil wars and revolutions which upset world markets, trade agreements and international political alliances.

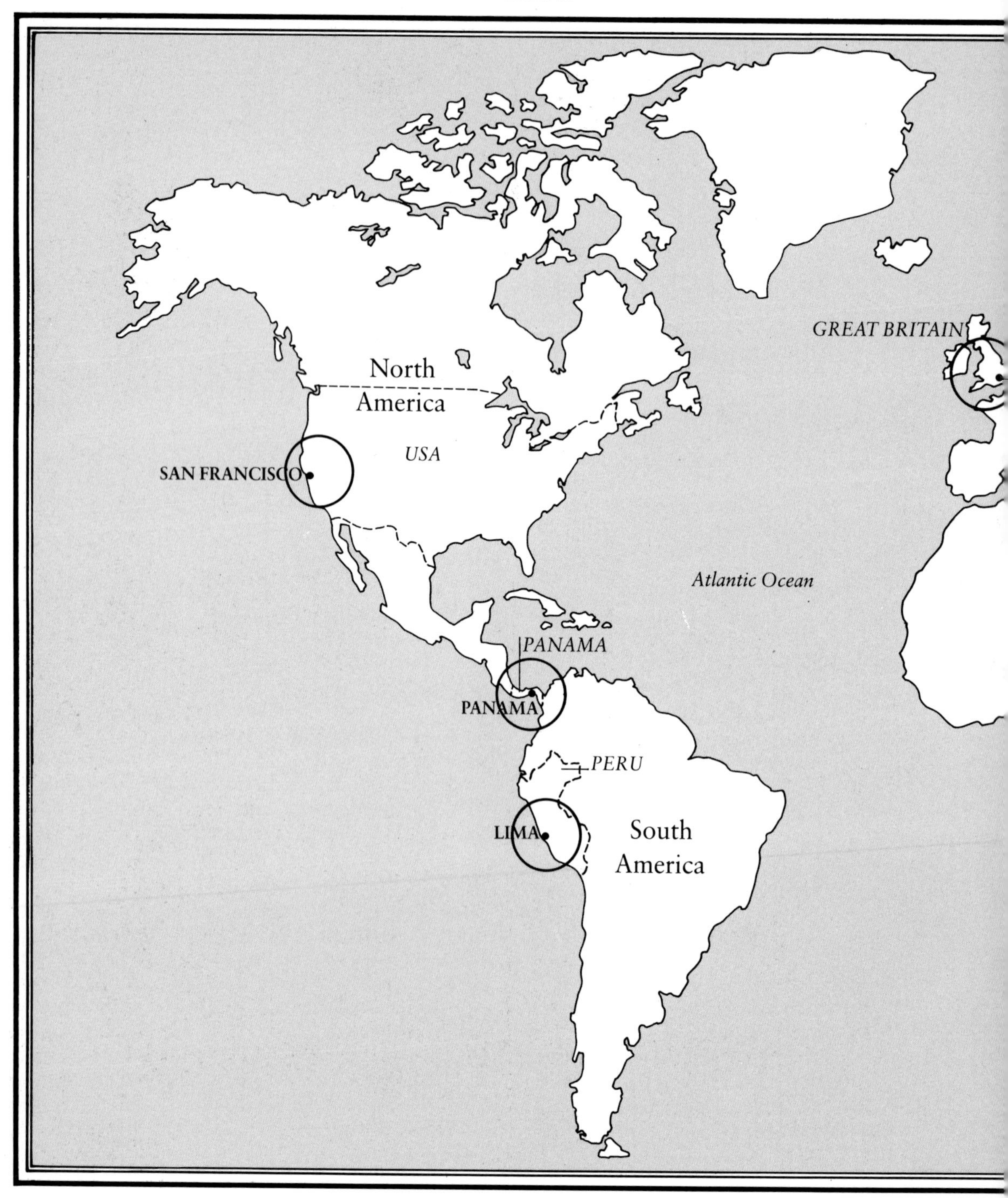
GREAT BRITAIN
North
America
USA
SAN FRANCISCO
Atlantic Ocean
PANAMA
PANAMA
PERU
LIMA
South
America

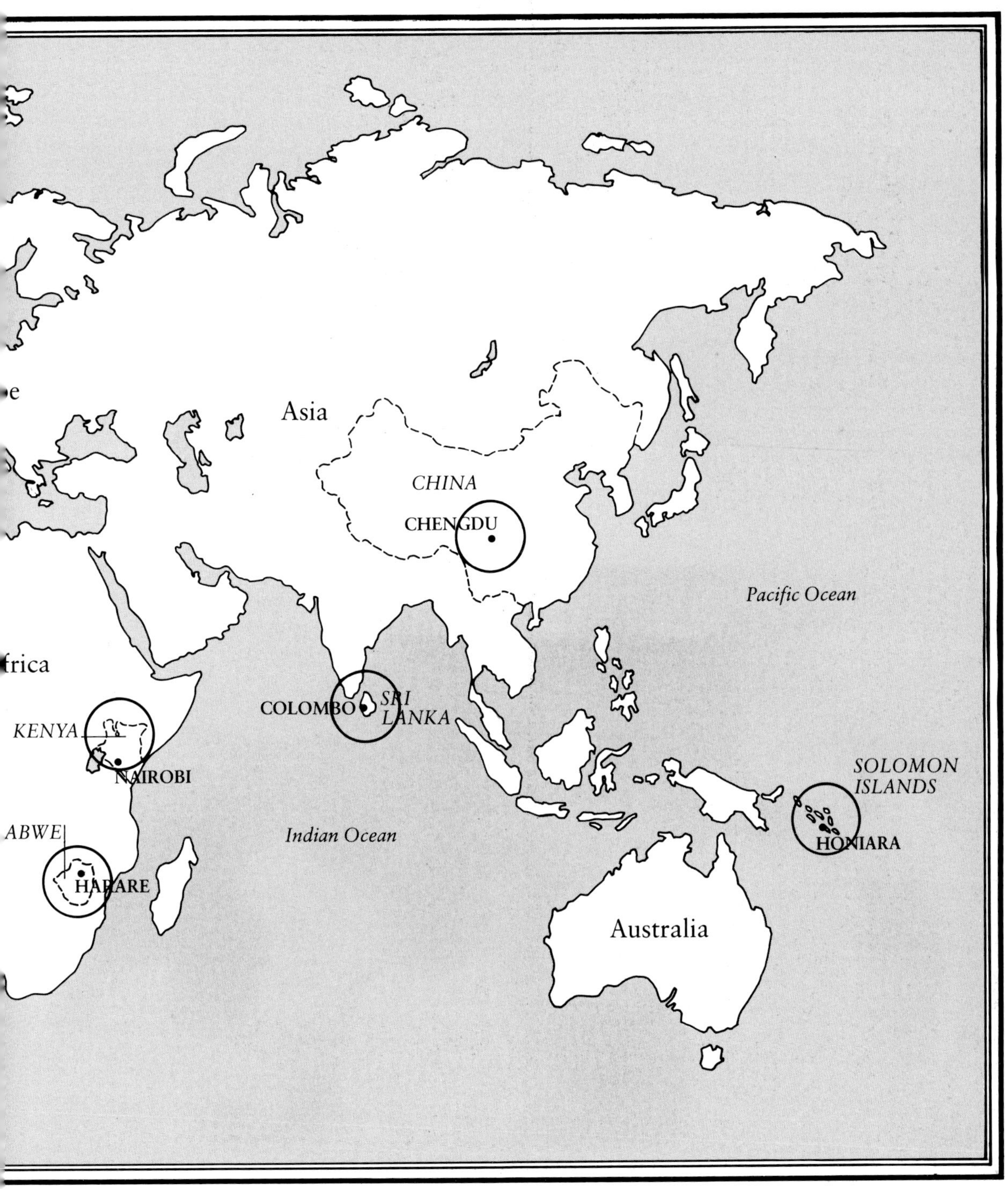

Asia
CHINA
CHENGDU
Pacific Ocean
frica
KENYA
NAIROBI
COLOMBO
SRI LANKA
SOLOMON ISLANDS
HONIARA
ABWE
HARARE
Indian Ocean
Australia

For these reasons – and because of foreign policy goals of the Northern nations – most governments in the North have established organisations designed to help the development of the lesser developed nations. Britain has its Overseas Development Administration (ODA), the USA its Agency for International Development (USAID), Sweden its International Development Authority (SIDA), the United Nations its UN Development Programme (UNDP), and so on. The International Bank for Reconstruction and Development, more widely known as the World Bank, was established in 1945 to provide capital to help nations rebuild after the Second World War and to finance development.

Although a lot of effort and many words have been spent in the attempt, the term 'development' has never really been defined to anyone's satisfaction. Most citizens of the North, if they think about it at all, tend to see development as a way of making Southerners more like Northerners, making them more 'modern'. They will have more disposable income; thus they will have enough food to eat; they will be healthier and live longer; they will live in sounder houses, cook with gas or electricity, have safer water supplies and proper plumbing; they will be able to dress better, and to move about more easily by road and rail. In short, the South will 'catch up' with the North. This is a goal proclaimed in speeches and declarations by both Northern and Southern political leaders. In some nations, it is actually coming to fruition for a few people.

In terms of world statistics, however, this kind of development is not taking place. There are more poor and hungry people today than at any time since the human species evolved. In 1980, some 730 million people in eighty-seven developing nations were not

A solar-powered television in Northern Niger brings in agricultural advice and Western soap operas

receiving sufficient food to enable them to lead active working lives. A decade earlier, the figure was slightly over 600 million. (These World Bank figures disregard China and the many who go hungry in the North.) Although during the 1970s the percentage of hungry people in these nations had actually fallen from 40 to 34 per cent, population growth overwhelmed percentage improvements in the food availability.

Today more people than ever before live in traditional huts or urban shanties, and over this and the next decade the world's growing populations will need half a billion new houses. It is unlikely that many of these structures will be safely and soundly built; on the contrary, most will be hastily erected by their occupants from discarded plastic, tin and cardboard. Each year more people are without safe water supplies. In 1975, 1·23 billion people in the South (again excluding China) were without clean water; by 1980, the figure had increased to 1·32 billion. Today an unprecedented number of people rely on wood, dung and straw for cooking and heating their homes; and an estimated 2·4 billion people will be running short of firewood by the end of this century.

A Calcutta family living in concrete pipes

Statistics such as these are not the main message of this book, which is devoted instead to signs of hope; but the signs of hope found in the coming chapters must be read against this backdrop of failure and despair.

As the national and international development agencies became aware of the complexity of their tasks, they seemed more and more to limit their activities to narrow, achievable and somewhat selfish goals. For instance, 80 per cent of Britain's nation-to-nation aid is now 'tied' to the purchase of British goods; that is, in order to receive the aid the recipient nations must buy British products and services, regardless of whether these products and services are appropriate. Similarly, Canada ties 80 per cent of its nation-to-nation aid to the purchase of Canadian goods. Development aid is used by many nations more as a tool of foreign policy than as a tool of development. Approximately one-third of all US bilateral aid in 1985 was earmarked for Israel and Egypt, largely because of the strategic importance of these nations. Nearly one-quarter of all aid from North to South is given to 'upper-middle-income nations', mainly for political reasons. There are profound debates over whether aid realistically could or should be handled otherwise, but taxpayers in the North are beginning to realise that the small proportion of their taxes which is apparently designated for 'development' is not in fact achieving that end.

Taxpayers in the North are also beginning to realise that many of their governments' policies are actually blocking development in the South. Northern-dominated organisations, such as the World Bank and the International Monetary Fund, have encouraged Southern agricultural nations to grow 'cash crops' or 'export crops' in order to earn hard currency. It is argued, not unreasonably, that this is the only way in which these poor farming nations can buy energy, medicines, books and equipment. Yet Northern governments then erect trade barriers in the form of quotas or tariffs on Southern commodities which they also grow themselves. The sugar policies of the European Common Market are estimated to cost Southern producers £1·3 billion ($2 billion) per year in lost earnings. Many of the drought-stricken African nations grow peanuts; however, they have a difficult time exporting these to the USA because of quotas imposed by the US government.

It is not only the Northern countries which are to blame: few Southern governments have done well in taking care of their poor majorities. The governments of many Southern agricultural nations invest only a small proportion of their revenue in agriculture, in spite of the fact that the majority of their people live by farming. These governments, often encouraged by Northern advisers, have concentrated their efforts and resources on industry and urban areas rather than on agriculture and the countryside. 'Almost all have failed to learn the real lessons of the development in North America and Europe,' said US economist J. K. Galbraith in 1985. Encouraging the South to build steel mills rather than help their farmers to grow crops 'is next only to our failure to perceive and act on the consequences of nuclear conflict, the most compelling error in social perception in our age', he added.

The USA during the nineteenth century, Europe before the industrial revolution and Imperial Russia all emphasised agriculture as the mainstay of their economies. But, 'Whenever the question of design for economic development in the new countries comes up, all the developed industrial countries look at their present industry, not at their past concern for agriculture, as their guiding example,' Galbraith argued. The development of farming puts schools in

Ethiopia once had 40 per cent forest; now it has 2 per cent. These seedlings will help hold the soil in place.

the countryside, builds systems of rural transport and encourages the rational and equitable use of land, he said. It also produces a rural population which is able to buy the goods produced by new industry. This was true when the USA was an agricultural nation; it is still true today when the great majority of people in the South work on the land.

Southern governments do not completely ignore the development of agriculture, but they tend to invest little in the area in which help is most needed – in the attempts of small farmers to grow food crops. Instead, governments may concentrate on small, highly productive areas, as in the case of India's 'Green Revolution' in the northern regions, where high-yield wheat and rice varieties are grown with expensive helpings of fertilisers, pesticides, herbicides and irrigation. Yet millions of Indian farmers can afford neither this technology nor the food it produces. Or the Southern governments may concentrate on one or two cash crops designed for export rather than for local consumption, as happens in so many African nations. Only about 6 per cent of Ethiopia's agricultural industry can be described as 'modern' and almost all of this grows coffee rather than food. Thus the efforts made by Southern governments in order to 'catch up' often assure that more and more of their growing populations are left further and further behind, and that many are left with little or no role in their nation's economic life.

The Swedish writer on Third World development, Jan Myrdal, offered in a 1985 speech a scathing personal view of the end results of this sort of unsustainable development: 'Science, industry and new capitalist agriculture based on the Green Revolution might develop. The administration might be efficient. But if 400 million poor Indian peasants and slum dwellers [out of a national population of 750 million] this very minute disappeared from the face of the Earth it would not make any difference to that economy. And the ruling class of India would not turn a hair – as long as the dead millions did not lie around stinking but disappeared in a clean fashion.'

There is a growing fear among development planners, both North and South, that development is not only not happening, but that many of our official development aid organisations and policies may be doing more harm than good, both to humans and to the land. The encouragement of cash crops has spread little wealth among the poor in the poor nations. The rich nations grow richer; the poor either grow richer much more slowly or, as in the case of many African nations, actually become poorer. Each year the gap widens: 'At the end of three decades of international action devoted to development the result, by 1985, is likely to be an increase of $50 (£33) per capita in the annual incomes of the poorest group over 1965 incomes, compared with an increase of $3900 (£2600) per capita for those of the richest,' according to Sir Shridath Ramphal, secretary-general of the Commonwealth Secretariat.

The land, upon which six-tenths of the world's population relies for its direct income and upon which the entire planet's inhabitants rely for their food and water, is also growing poorer. In many cases, efforts to develop a given nation are destroying the very resources upon which that nation's future development must be based. Each year large tracts of once productive soil are being eroded, degraded or turned to waterlogged, salty or desert-like landscapes in almost every nation of Latin America, Africa and South-east Asia. These are not the findings of radical environmental groups, but are facts revealed in the national *Environmental Profiles* of developing countries being compiled by the US Agency for International Development.

The efforts made by poor farmers to stay alive and by poor nations to grow more crops for export are destroying farmland in what are essentially agricultural nations. According to UN studies, one-fifth of the planet's surface, over half of all of its arid and semi-arid land, is under direct threat of becoming worthless desert. The tropical forests which protect much tropical farmland from erosion and which house half the world's species of living plants and animals are vanishing as ever more forest is cleared; a swatch the size of Ireland disappears each year.

This destruction not only speeds erosion and denies humans fuel for living, but it wipes out plant species which could provide future food crops, medicines and industrial chemicals. Almost half of our drugs and pharmaceuticals are based on the chemistry of wild plants; one obscure tropical forest plant, the rosy periwinkle, has saved the lives of countless Northern children suffering from leukaemia. But many plants vanish before they can be studied by scientists. No one can be sure, but some experts estimate that every day another species becomes extinct. By the year 2000, we could be losing 130 species per day as the destruction of wild land accelerates.

These environmental disasters need not be slow and steady; they can come about suddenly. In the early 1980s, the Swedish Red Cross began to wonder why each year it was receiving more and more appeals for relief due to natural disasters in the Third World. Joint studies undertaken by the Swedish Red Cross and the international environment and development information service Earthscan sought the reasons, and found that people's attempts to survive were actually causing many of the large and growing number of the disasters usually described as 'natural'. The disasters were directly related to failures in development.

Drought is the catastrophe which affects more people in the world than any other, and its carnage is increasing with each passing decade. The studies show that during the 1960s drought struck 18·5 million people every year; by the 1970s that figure had climbed to 24·4 million people annually. These studies were completed before the African drought and famines of 1984–5, which affected some 30 million people on one continent alone, and before the strangely less-publicised Indian droughts of 1985–6, which afflicted 100 million people in nine states. Drought has to do not only with annual rainfall, but with how much of the water which does fall is absorbed and held by the soil. Clearing the land of trees and bushes causes water to run off the soil quickly, making less water available for humans and their crops. The African and Indian droughts were thus directly related to the ways in which people had been using their land.

Floods, although they affect fewer people yearly, are increasing in impact even faster than droughts: in the 1960s 5·2 million people were affected by floods each year; by the 1970s the number had risen to 15·4 million. The trend has continued into the 1980s, in Bangladesh, South-east Asia and the Andean nations of Latin America. The cause is the same: stripping the land of vegetation means that water runs off more quickly; rivers cannot absorb the excess, and in rainy areas the result is more, and fiercer, floods. The studies even found that the surprising growth in the numbers of victims of such disasters as earthquakes and tropical cyclones had much to do with the increasing numbers of poor people living in poor housing in dangerous areas exposed to earthshock and storm surges.

The Mexico City earthquake, 1985. Poor people in poor nations are most vulnerable to 'natural' disasters.

The research also revealed that disasters which take heavy tolls in the South are not so deadly in the North. In the two decades from 1960 to 1980, Japan suffered forty-three 'natural' disasters of various types and lost an average of sixty-three people per disaster; Peru suffered thirty-one such disasters and lost an average of 2900 people in each event. The Southern nations and their people become more vulnerable each year. They are part of a vicious cycle:

poverty forces nations and people to misuse their land; the misused land in turn makes people and nations poorer and more vulnerable on land which is more prone to disasters.

Taken together, these trends – the growing number of poor and hungry people, the decline in environmental resources for the future, and the rapid rise in the number and scale of environmentally-linked disasters – have caused growing disenchantment with development as it is currently practised. The search is taking place for a new approach, a new model. The concept most favoured is usually summed up under the vague term 'sustainable development'. It is not a particularly exciting, or dramatic, concept, any more than is the concept of 'living within one's budget'. As with 'development', there are problems in definition. One can easily define and point a finger at examples of 'unsustainable development'; one need only look at the declining food yields, the droughts and the spread of deserts in much of dryland Africa. It is much harder to find examples of development schemes which make large numbers of poor people better off through processes which can be sustained over many years, development which makes such people less vulnerable to changes in weather, in commodity prices and in trade barriers.

Given the scarcity of concrete examples, one is usually forced to discuss sustainable development in terms of generalities. For instance, it is *stable*; it does not disrupt ecological systems or over-exploit natural resources, as is the case with the livestock projects in Panama (see Chapter Two). Neither does it disrupt culture and societies, as do the rapid migrations of farmers from the Peruvian countryside into Lima (see Chapter Six). On the contrary, this stability will involve the rational use of renewable resources, keeping options open for future generations who may have a clearer understanding and appreciation than we do today of the value of such resources.

It is more *resilient* to change, either in weather or governments. The technology used in the sustainable development package is *appropriate* to the needs, skills, training and finances of the people using it, as are the tuna fishing techniques of the Solomon Islanders (see Chapter Five). The most often repeated mistake of much development work is that efforts have been based on technology which has little to do with the needs of the people, as with the deep wells in Kenya (see Chapter Four).

Sustainable development is *productive*. It creates surpluses above the needs for minimal survival, as does the new form of livestock production and marketing among the Rendille tribespeople of Northern Kenya (see Chapter Four) and the 'Responsibility System' being tried out in rural China (see Chapter Seven). It is *self-reliant*, being based on the efforts and ideals of the developing peoples and reflecting their needs, as in the case of the Sri Lankan village described in Chapter Three. It does not rely solely on the knowledge of outside experts, although such expertise, guided by local people, can be an important factor in the success of development work.

As shown by the examination of British agriculture (see Chapter Eight) and drinking-water pollution in California (see Chapter Nine), there is still much need of sustainable development in the North, where many human practices limit the options of future generations. (Beyond the scope of this book are such patently short-sighted activities as acid rain pollution, the warming of the planet through the release into the air of large quantities of carbon dioxide, and that most unsustainable of all human activities, the proliferation of potentially planet-destroying nuclear weapons.)

Another important aspect of sustainable development is that it is more complicated and challenging than the old-fashioned sort. British development expert Robert Chambers argues that increasing the yields and improving the prospects of many small farmers who farm by intricate mixes of crops and animals on small tracts of land is a much more complex endeavour than covering those farmers' fields with one huge, high-technology plantation growing a single crop.

Thus the snapshots of sustainable development in this book, taken largely through the eyes of local people, are intricate and detailed. The answer is never found in a simple solution such as the introduction of a new crop or the building of a clinic. The Rendille herders of Kenya needed wells, veterinary services, protection from armed stock raiders, new land-use strategies, access to markets, education in marketing their product and shops in which to spend their profits. Nor can it be guaranteed that any of the

Oasis irrigation, Western Sudan. Traditional systems often work better than imported technology.

efforts outlined in the following chapters will ultimately succeed. Each could so easily be overwhelmed by local events or by economic and political decisions taken in a nation's own capital or in the capitals of the more powerful nations of the North.

Is the hoped-for result even possible? Many people in the world today are deeply pessimistic about our ability to manage our planet. The British development expert Piers Blaikie made a study of the political and economic reasons for the widespread soil erosion in the South. He concluded that this erosion could never be controlled unless the wealthy and powerful nations of the planet saw it as a direct threat to their livelihoods and life-styles. But, as things are, these are the very peoples who are farthest removed and most insulated from the effects of soil erosion. They are also the least affected by deforestation, pollution, the extinction of wild species and other forms of environmental destruction. Blaikie concluded his study with the mild understatement: 'Thus the prospect for effective soil conservation remains rather poor.'

Other writers outside the development business are also pessimistic. The American novelist and travel writer Paul Theroux watched from the deck of a boat on China's Yangtze River as thousands of peasants toiled in the mud. He saw a vision of the entire planet's future:

> Forget rocket-ships, super-technology, moving sidewalks and all the rubbishy hope of science fiction ... We are stuck with this mildly poisoned planet and its smoky air. We are in for hunger and hard work, the highest stage of poverty. No starvation, but crudeness everywhere; clumsy art, simple language, bad books, brutal laws, plain vegetables and clothes of one colour. It will be damp and dull, like this. It will be monochrome and crowded – how could it be different? There will be no star wars or galactic empire and no more money to waste on loony nationalism in space programmes. Our grandchildren will probably live in a version of China. On the dark brown banks of the Yangtze the future has already arrived.

There have been a number of such visions, some from more academic sources. Perhaps the best known is the Club of Rome's *Limits of Growth* report of the early 1970s, which described the collapse of systems as the planet runs short of resources. Since *Limits of Growth*, many critics of such forecasts have argued forcefully that humans are endlessly adaptable creatures which are able to tinker and change their world systems as conditions demand.

This book does not take sides in that debate (although it does offer a view of China which is different from that given by Theroux). It argues that many of the environmental systems upon which life on the planet depends are under great human pressure, and that it is long past the time for humans, especially those who guide national and international policies, to begin tinkering and changing if we are to adapt to environmental realities. More important, it offers examples of how and where efforts are already being made, by ordinary people, to find ways of living and even thriving within their own means and within the means of their only planet.

II

FROM WASHINGTON TO PANAMA: BUYING DESTRUCTION

'Deforestation may be being financed with our money, but it is very much against our philosophy.'
Inter-American Development Bank official

A few miles inland from the Caribbean north coast of Panama lies the farm of Arnulfo Corte, set amid rolling hills which were once covered with trees. Corte is an honest, hard-working cattle rancher who rears a herd of 160 cows on poor pastureland. He is not rich by Northern standards, but he is much better off in terms of cash and land than are the vast majority of people in Latin America. He is taking part in the 'development' of his nation, producing beef for export in order to help Panama pay its huge national debt.

Not far from the White House in Washington DC lies the headquarters of the Inter-American Development Bank, an imposing office building with a foyer filled with potted plants and trees and multi-tiered fountains. The entrance to the bank seems to be more lush and green and better watered than Arnulfo's farm. The bank is devoted to the 'development' of the states of Latin America; its aim is to help those nations and their peoples towards a richer, more secure future.

The connection between the farm and the bank is a relatively small amount of money: £8670 ($13,000). The BBC television producer Brian Leith recently followed that sum all the way from the bank's main office on New York Avenue to Arnulfo's farm far off the main roads in Panama. Leith discovered that Arnulfo was not using the money in the way specified by the Inter-American Development Bank's documents. He found that, instead of financing development to make the world a better place, the money was financing destruction which would rob future generations of their chance of a reasonable livelihood. He found that much larger sums were all being used for the same destructive purposes.

Arnulfo Corte

Before Leith began his international paper-chase, however, he examined the international development scene in Washington, and in particular the controversy which has grown up around the environmental effects of the actions of the international development banks. He first visited the offices of the organisation best known by its official nickname of the 'World Bank'. Its real name is the 'International

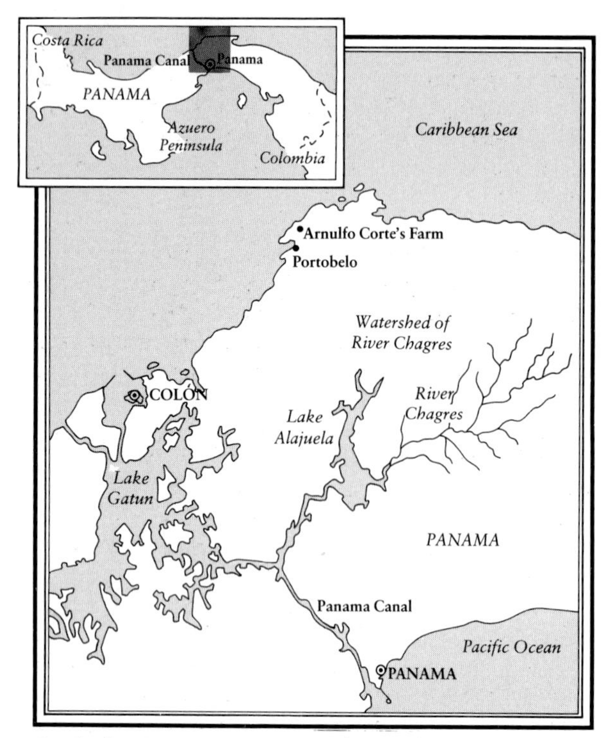

Bank for Reconstruction and Development', which gives a better idea of its purposes and goals. It employs a professional staff of some 2400, 95 per cent of whom work in the Washington headquarters. It is the world's principal lending institution for development capital and, given the size and credentials of its staff, it is the world's centre for expertise on development matters. It has 134 member countries but, unlike the United Nations with its one-nation, one-vote rule, voting power at the World Bank depends on the size of a nation's paid subscriptions to the Bank, subscriptions being assessed on a nation's ability to pay, with the richer nations putting in more. As the Northern nations pay more, they have more say in the Bank's polices.

Recently, the World Bank has come under fire from several quarters – academics, grassroots US environmental pressure groups and the US Congress – for funding projects which appear to be destroying both the local environment and local cultures. A worldwide day of demonstration and marches in protest against the World Bank and other development banks was organised in late September 1986. Concern over the Bank's policies is not only voiced in Northern countries, and it is not always peacefully expressed; in June 1986 a crowd of demonstrators surrounded and burned to the ground a World Bank-financed factory being built on Thailand's scenic Phuket Island. The £30 million ($45 million) factory, a refinery for the production of the metal tantalum, which is used in items ranging from light-bulbs to nuclear reactor components, was designed to aid the economic development of the area. The demonstrators feared that the factory would release hydrofluoric acid or radioactive pollution, or both, into the sea, thereby destroying the tourist trade and the fisheries. They saw not development, but disaster.

One World Bank scheme which united several environmental groups in protest was a plan to develop north-eastern Brazil's vast Polonoroeste region, three-quarters of the size of France and until recently covered in largely untouched forest. The scheme involved the building through the Amazonian jungle of a 1000-mile road which brought in new settlers at the rate of 13,000 per month. The settlers, few of whom had any experience of farming jungle,

Slash and burn clearing of rainforest in Panama

cut down trees, invaded Indian lands, and were left with infertile soil which was unable to support the crops they planted. The promise of the establishment of national parks was not fulfilled, and forest which was to have been left standing was wiped out. Environmental groups testifying before the US House of Representatives' Banking Subcommittee on International Finance suggested that at the then rate of deforestation the entire region would be a desert within a few years.

Staff director Barry Hager voiced the subcommittee's concern that in the Bank's 'rush to develop and create economic wealth, in some cases environmental damage is done which is inconsistent with long-term development'. After hearing evidence on the Brazilian scheme and other controversial projects in Indonesia, India and southern Africa, the subcommittee produced guidelines for the World Bank and other international development banks which receive US funding. The World Bank itself, just a few days before the Phuket incident, had announced new conservation standards for its projects and promised the creation of a global database, developed with the help of satellites, on environ-

mental conditions in developing nat[illegible] froze £170 million ($255 million) i[illegible] Brazilian project. 'This shows that th[illegible] environmental and tribal issues very s[illegible] after a long delay,' said Dr Robert C[illegible] of the Bank's environmental office.

The Inter-American Development Bank, a smaller version of the World Bank, is one of the so-called regional development banks (which include the African Development Bank and the Asian Development Bank) specialising in a given area of the globe. It has forty-one members, most of them in North and South America but a few from Europe. These regional banks have also been criticised for financing projects which may be doing more harm than good, and have recently promised to do better. The Inter-American Development Bank has pulled out of a scheme similar to the Polonoroeste project in the Brazilian state of Acre.

Such criticism puts the development banks under a pressure which they do not regard as entirely fair, because they do not simply invent projects and loans at headquarters. They claim to base their activities on the aspirations of the nations on the receiving end of the money. A World Bank spokesman said that it would have been far easier to stay out of Polonoroeste altogether, but the Bank's 'playing it safe' would not have prevented environmental destruction and the invasion of Indian territory.

Given the controversy, the criticism and the development banks' claims to have become more environmentally sensitive, Leith thought it would be revealing to pick, more or less at random, a project which might be suspected of harming the environment, and then to follow the trail of decision-making and the issue of money from Washington to the project site itself. Leith is attached to the BBC's Natural History Unit, and did not go into his investigations with any preconceived notions about how development works, or should work. He wanted to see for himself.

Livestock projects seemed to be likely candidates, especially those in heavily forested tropical nations, because in such terrain it is usually necessary to cut down forestry to make room for the livestock. Given the right conditions, this is a sensible thing to do. The soil under the trees must be both fertile enough well enough protected from the erosion caused by tropical rains to support pastures over the long term, and the farmers concerned must have the skill and resources necessary for the management of land and cattle.

Both the World Bank and the Inter-American Development Bank have been loaning money to Panama to improve that country's cattle ranching industry. The World Bank's 'Livestock Project Number Three' involves about £6 million ($9 million) in finance and runs from 1984 to 1990. The Inter-American Development Bank is funding a more concentrated 'Livestock Development Programme' in Panama to the tune of £6·7 million ($10 million), which was just coming to an end in 1986. The documentation on both projects claims that the loans are designed not only to increase production of beef, but to improve existing pastures, to replant trees in deforested areas, and to encourage sensible and sustainable use of the land. These goals are clearly set down in writing as part of the banks' official policy.

Leith found it difficult to evaluate the World Bank's project, both because it was just beginning and because the World Bank tends to be very secretive about the effects of its projects until long after they have ended. In fact, even after a scheme has ended, the World Bank usually refers questions on it to the government of the nation involved. Besides, the World Bank has no office or permanent staff member in Panama, the country where the money was to be spent.

However, the Inter-American Development Bank, focusing as it does more closely on Latin America, is well represented in Panama. This proved fortunate for Leith, because the staff at the Washington headquarters did not want to talk to him about Panama's Livestock Development Programme. Abraham Arce, head of the bank's Central American Desk, suggested that he see John Pino, a consultant who had advised on a previous livestock project. Pino said that he was not in a position to give details as he was not an official spokesman for the bank. However, Steve McGaughey, a forest economist at the bank, was more forthcoming. He admitted to Leith, without going into details, that the programme raised certain environmental problems, of which the bank was aware and with which it was trying to cope. But 'We try to do what the host nation wants and we cannot interfere in its internal affairs. The bank serves the perceived needs of the host governments and the priorities as seen by those governments,' McGaughey said. He could not make documents on the programme available as these were not meant for public viewing. He suggested that Leith go to Panama.

Panama City, on the Pacific end of the Panama Canal, is a modern city with a large number of offices of banks and trading companies which have been attracted by the nation's plans to become a world banking and free-trade centre. The Panamanian banking secrecy rules, similar to those in Switzerland, and the exemption of offshore loan operations from taxation have been effective in encouraging foreign capital investment, as has the use of US currency. The top floor of one of the the most modern office blocks houses the Panamanian offices of the Inter-American Development Bank; there the local executives were much more helpful than those in Washington. They knew that Leith had been talking to their superiors at head office, and they wanted to present their work in the best possible light, both to Leith and to Washington. An official explained that both the World Bank and the Inter-American Bank worked through host national development banks, the Inter-American Development Bank having allied itself with the Banco de Desarrollo Agropecuario (the Bank for the Development of Agriculture and Livestock, known locally as the BDA), and that this was where the Inter-American Development Bank's money for livestock had gone. He suggested that Leith talk to the people at the BDA.

Panama has been keen recently to develop its agriculture and livestock because it has been having trouble repaying its debts, which on a per capita basis are bigger than those of Brazil, Mexico and most other Latin American nations. In order to help the nation raise money for the debt repayments, the International Monetary Fund (IMF) had given Panama loans so that it could sharply increase its exports of beef, coffee, sugar, shrimps, bananas and vegetable oil. However, as Panama has concentrated its efforts on developing banking and trade in and around the Canal Zone, its agriculture has declined so that, whereas twenty-five years ago agriculture represented almost a quarter of the Gross Domestic

Panama City. Banks and free-trade zones attract wealth to the city but rural people make their land poorer.

Product, it now provides just above one-tenth of the GDP. The farmers in the countryside, where 45 per cent of the population lives, have suffered as a result. For those living in the city the average income is £1530 ($2300) a year, but for those in the countryside the average income is only £465 ($700).

The vast interior to the west of the Canal has been deforested by farmers known locally as *arreras*, a name which is taken from the Spanish for the leaf-cutter ants which bite off and carry away leaves from which they make their nests. The human *arreras* are forest cutters and burners who clear land for pasture and crops. As in much of Latin America, the rich in Panama want to be ranchers, regardless of the jobs they may have in the city; they want to ride horses across their spreads and to have their own brand on their own cattle. The image of the rich *ranchero*, the man on horseback, is embedded deep in the colonial culture; and the forest is the enemy of the rancher. (Throughout much of Latin America, the people's deep-seated hostility to natural forests is reflected in their vocabulary: natural vegetation is referred to as *monte* (untamed forest), while the term *bosque* (woodland) is reserved mainly for man-made tree plantations.)

The untamed forest is there to be tamed; that is, cut down. In the late 1960s, the former Panamanian leader General Omar Torrijos launched his *La Conquista del Atlántico* (The Conquest of the Atlantic) programme, which undertook to clear the forests on the Atlantic slopes of the continental divide. Today, the only way in which peasants are able to gain the use of land owned by the ranchers is to clear it for pasture. They are allowed to plant crops for a few years while the land is still fertile, but eventually most of it reverts back to the big landowners, who use it for herding cattle.

The soil under most rainforest is low in fertility and high in minerals such as iron and aluminium; the work of the *arreras* bared these very poor soils to the rain and sun, causing erosion and the spread of worthless, rock-hard land over large areas of the interior. The Sarigua Desert on the eastern coast of the Azuero Peninsula, which juts southwards into the Pacific, is slowly spreading inland. Peasants are being squeezed out of the interior both by erosion and by the needs of the wealthy ranchers for more and more land to support their herds. One can sit beside the Pan American Highway to the east of the Canal and watch busloads of these peasants, these *gente del interior* (people of the interior), riding toward the forests which extend to the borders of Colombia. Once there, they clear the land by cutting down more forests, and so the process starts again.

These are the conditions under which the BDA and the Panamanian government are trying to improve livestock production for the export and domestic consumption of beef. Such grass-fed beef is suitable for the manufacture of hamburgers, so there is a direct link between the destruction of the forests in Central and Latin America and the desire of Northerners to eat cheap hamburgers. This link is often referred to as 'the hamburger connection'.

At the local BDA office Leith found the executives responsible for livestock to be frank and helpful. But, as they rarely went out into the countryside, they had no clear idea of how the money was actually being spent on the farms. They therefore arranged for Leith to meet a bright and ambitious young Panamanian named Benigno Jarmillo, a BDA 'field engineer'. Jarmillo is the man responsible for deciding which farmers in a given area actually receive loans, thus becoming the ultimate beneficiaries of the money which originated in Washington at the Inter-American Development Bank. Jarmillo's territory lies to the north and east of the Atlantic end of the Canal, an area dominated by the village of Portobelo, the burial place of the Elizabethan Francis Drake, and once one of the main Spanish ports of the Caribbean. The village itself comprises a collection of tin huts, dominated by the massive ruins of the stone fort. It also has a local BDA office.

Jarmillo, Leith and the local BDA official responsible for collecting repayments travelled by jeep down rutted dirt roads to the farm of Arnulfo Corte, who had a loan of £8670 ($13,000) outstanding with the BDA. He was paying 9 per cent interest, a few points below commercial rates. Leith found it interesting that the BDA, rather than himself, had chosen Corte's farm upon which to focus. The BDA apparently picked Corte because it was proud of what he had done to his land and regarded him as one of the programme's model farmers and star attractions.

The farm covers almost 250 acres of hilly land. The jeep had to be abandoned a mile from the farmhouse, and the visitors removed their shoes in order to wade through two streams. The farmhouse itself is a small hut with a banana-leaf thatch roof. The walls are made from loosely constructed planks between which the sunlight streams. Corte himself is a big, raw-boned man with the high cheekbones typical of the *mestizos* – those with mixed Spanish and Indian blood – who comprise 70 per cent of Panama's population. He appears diligent, energetic and tough; the skin on his hands is as hard as the boards of his house.

In 1985, Corte owned a herd of 111 cattle. The next year he had increased this number to 160 by taking care of his calves, a prime example of his skill and hard work and another reason why the BDA chose his farm as a showcase. Eleven years earlier, he had come from the eroded interior west of the Canal in order to clear this land of forest. Thus he was much better established than most of his neighbours, many of whom also have BDA loans, who have been in the area for only a few years and were still clearing their land. Most of them were also *gente del interior*.

Corte's farm lies in the middle of the vast Portobelo Buffer Park, a national 'forest park' which is designed to create a buffer zone between farmers and the Upper Chagres National Park along the Chagres River which feeds water into the Canal, the linchpin of Panama's economy. His land is supposedly controlled by RENARE, the national department of natural resources. He had cleared all but fifty acres, and planned to deal with those soon, though he would leave stands of trees for building materials and firewood.

Leith found himself becoming rather confused. He was in a forest park designed to help protect one of the world's most important waterways. He was on land legally controlled by those responsible for preserving the nation's natural resources. Yet he was talking to an honest, hard-working farmer who was clearing the trees from that land. Not only was Corte not breaking any laws, but he had the encouragement of his government and was using capital provided by one of his nation's development banks and by the Inter-American Development Bank in Washington.

Arnulfo Corte and his herd

Many of the cows on Corte's land actually bore the brand of the BDA, rather than the farmer's personal brand.

Corte's farm lies just north of the Panama Canal watershed – that is, the area from which any rainfall eventually flows via streams and rivers into the Canal. But the farms of many of his neighbours, who were also receiving BDA loans, were situated well within the Canal watershed; in fact some of them were sited on the shores of key lakes and waterways which feed the Canal directly. Leith found no evidence of BDA, or any other, official restrictions on deforestation; he found only encouragement for the felling of trees. He discovered that money from the World Bank 'livestock improvement' project was also going to farmers in the watershed.

Deforestation in this case is not primarily concerned with the issues of loss of wild species or of beautiful ecosystems. These forests are vital to the continuing functioning of the 51-mile-long Canal, which is basically a series of locks at both ends with the large, man-made Lake Gatun in the middle. The ships passing through the Canal must be raised eighty-five feet above sea-level via a series of steps – the locks – over the continental divide and down on the other side. The raising and lowering of the water in the locks to allow one ship to travel through the Canal costs about 40 million gallons of fresh water, and thirty-two ships go through each day. The same water system provides electricity and drinking and industrial water for the big cities of Panama City and Colón at opposite ends of the Canal. All of this

Panama Canal. Erosion from its rapidly deforested watershed brings more silt into the canal each year.

requires an estimated 2 to 3 billion gallons per day. Much of this water is brought into the lake by the Chagres River, which is dammed higher up to form Lake Alajuela (formerly Lake Madden).

The only way in which the Canal's water can be replenished is by rainfall throughout its watershed. If the watershed is forested, the water seeps into the ground or runs slowly into the streams. Water reaches the Canal steadily, all the year round. The deforestation of the watershed causes rainfall to rush into the key lakes and rivers, carrying a lot of soil and silting up the reservoirs in the process. In 1935 it was estimated that it would take 200 to 300 years for Lake Gatun to silt up 5 per cent. But that 5 per cent point has now been reached. Lake Alajuela is a deep red colour, the same colour as the soil which banks its shores and which gushes into it with each rainfall. As the lake's shores are eroded one can almost watch new land being formed and grass taking root.

In addition, the deforestation seems to be having an unpredicted effect on the watershed. A study carried out by the Smithsonian Institute's Tropical Research Centre in Panama found that rainfall has been fairly constant over decades at meteorological stations at both ends of the Canal. But inland meteorological stations within the critical watershed zone have been showing a decline in rainfall of about one inch per year over a number of years, in an area which has an average rainfall of 80–120 inches per year. During the dry years of 1982–3, ships had to offload cargoes and send them by rail from one end of the Canal to the other so that they could travel safely through the lowered water levels in Lake Gatun and over the silt on the lake's bottom. Scientists are only now beginning to understand how forests recycle rainwater. It evaporates from the moist forests and is transpired from the leaves, to fall again as rain. Furthermore, cleared ground reflects more solar energy than does an area covered by a leafy canopy, and this energy interferes with the build-up of rainclouds. Whatever the precise mechanics, the Canal has been receiving less water each year.

Farmers are not only clearing the buffer forest park, but have been moving into and clearing the forests of the Upper Chagres Park, a forest which a Panamanian government environmental consultant described as 'perhaps the most critical forest in the world', given its function of protecting one of the world's key waterways.

As Leith was being driven at speed through this forest in a jeep, officials accompanying him were explaining how no farmers were being allowed to settle in this valuable area. Suddenly, Leith's vehicle almost crashed head on into a jeep coming fast in the opposite direction down the narrow road. The second jeep belonged to the IRHE, the Panamanian department of energy, and when Leith met the IRHE officials they freely admitted that they were in the area in order to supply electricity to the new homes of the farmers who had settled in the forest. It seemed to Leith that this was an odd way to discourage settlement.

In his conversations with BDA executives, with its field officials and with the farmers themselves, Leith drew nothing but blank stares when he asked about plans to replant trees on individual farms. The Panamanians were neither hostile nor suspicious, but were genuinely puzzled as to why farmers should either conserve or plant trees. 'There is no money to be made in planting trees,' was the universal response. References to management of the watershed and to the future of the Canal were lost on them. The Canal was outside their jurisdiction, not their responsibility. Their futures, whether they were farmers or BDA executives, depended on increasing beef production. This meant clearing trees for pasture.

Nor was there money to be made from restoring the eroded pastureland of the interior to the west. In fact, field engineer Jarmillo ridiculed the farmers in the interior because of their ignorance and the way they had ruined their land. He did not seem to realise that the bank for which he worked was financing precisely the same destruction near the Canal, on land such as Corte's. Corte and his neighbours, being mostly from the interior, were the very farmers who had ruined that land, the farmers for whom Jarmillo now had so little respect. Yet that flat interior land is less prone to erosion and is more suitable for sustainable pasture management than the hills inland from Portobelo. The farmers near the Canal lacked experience in managing hilly land and were taking no steps to prevent erosion.

Leith had followed the trail of the development money and found it to be a surreal odyssey, full of ironies but lacking in villains or scoundrels. The men he had met in Washington, Panama City, Portobelo and on the farms all felt that they were doing what was expected of them, and doing it well. The livestock programme had been devised by economists, bankers and agriculturalists to help to 'develop' a nation; yet it could in the relatively short space of a few decades play a significant role in crippling the nation's main industry, the Canal. The programme made perfect financial sense – provided the balance sheets did not include the eventual costs of the deforestation, erosion, desertification, siltation of the lakes and the cost to the farmers and the nation of their moving off the ruined land. The plans in Washington said that reforestation and pasture improvement were part of the package; Leith found no evidence of this on the farms in Panama. On the contrary, he found that officials in Washington, and even in Panama City, either did not know or could not bring themselves to admit publicly how the money was being used. Steve McGaughey, the forest economist at the Inter-American Development Bank headquarters in Washington, had been asked specifically, 'Does bank money go to finance deforestation?' He had replied, 'Deforestation may be being financed with our money, but it is very much against our philosophy.'

Leith wondered how any Congressional subcommittee guidelines, the World Bank's new conservation standards, and the planned satellite monitoring of the Third World environment could cope with the basic premise, in Panama and elsewhere, that cows and big farms represent development, while forests are enemies to be subdued.

The Panamanian project provides a dramatic, but hardly a unique, case. During the third quarter of this century alone, the area of man-made pastureland in Central America more than doubled, almost all at the expense of virgin moist tropical forests. (But beef consumption by Central Americans declined in the 1970s, as most of the meat was exported.) The syndrome extends into South America; between 1966 and 1978 some 31,000 square miles of Brazil's Amazonian jungle – an area equal in size to Austria – were cleared for 336 cattle ranches under the auspices of the Superintendency for Development of Amazonia, and another 20,000 ranches of varying sizes were also established. The government's goal was to make Brazil the world's leading beef exporter by the early 1980s. But by 1980 Brazil remained a net importer of beef because the pasturelands under the forests were not as fertile as had been expected.

Leading Brazilian environmental crusader Jose Lutzenberger noted that on Brazil's interior cattle ranches meat production hardly reaches forty-five pounds per acre, while on northern European farms which do not use imported feed, meat production is nearer 535 pounds per acre. Lutzenberger estimated that an acre of intact forest could produce about ten times more food, in the form of tropical fruit, nuts, game and fish, than could an acre of pasture. Of more concern in a job-hungry nation, the Brazilian cattle ranching industry employs only one worker per 2000 cattle, or one worker for every 7400 acres. Such figures led the US National Research Council to report in 1980 that 'the price of a US hamburger does not reflect total costs, and especially the environmental costs, of its production in Latin America'.

A depressing pattern emerges: so much destruction – destruction which will lead to loss of income and livelihoods over a relatively short term – appears to be financed by institutions with the word 'Development' in their titles. The Panama story is an indictment of the development banks involved. But the banks themselves are dealing with nations which have a blinkered view of development, nations which seem happy to destroy for short-term gains the resources which future generations, perhaps the very next generation, will need in order to survive. Should the banks be expected to impose their own standards on these nations? Banks are generally not opposed to imposing standards upon those to whom they lend money; they check collateral; they ascertain whether the money is being spent in financially sound ways. But the development banks have rarely, if ever, included environmental resources among the 'collateral' they consider.

But this argument misses a stage. Before the banks can even consider what to do as regards environmental standards, they must develop such standards. They have been extremely slow to do this. That may indeed be changing: 'Another irony in the story is that the only people I met who seemed to have both exciting and realistic ideas for resolving this dilemma and getting the balance sheets right were actually inside the World Bank,' Leith said. One of these people was a young British economist, Jerry Warford, a World Bank official who admits that he does not believe in 'conservation' for its own sake. Warford, and some other young economists inside and outside the development banks, believes that the only realistic way to base development on a sound, sensible, sustainable footing is through the grim science of economics. But he believes that development economics in the future must be radically restructured in order to take in the real costs; the long-term costs of lost soil, water and forests, the cost of moving people off the land, the cost of forcing people to move towards destitution in cities. Only then can 'development' really take place; over the long term which will make human life more or less difficult on this planet in the coming generations.

Vast slums in Mexico City. Each year, more Third World people live in such slums and shantytowns.

Queuing for water, Zimbabwe. Planners have been unable to keep up with growing populations' needs for clean water.

Floods in Peru, 1983. Deforestation of the world's mountain ranges brings more and bigger floods each year.

Amazonian Peru. The road brought settlers who cleared the hills and this caused a landslide which obliterated the road.

Auca Indians in an Ecuador rainforest (opposite). *Such forest dwellers are the only people on Earth who know how to manage a rainforest effectively.*

Auca land after a Western company cleared the forest to search for oil

Arnulfo Corte clears some of his few remaining woods (below and right). *The bare hills of his farm* (left) *will erode under tropical rains and bake hard under the tropical sun, supporting fewer cattle.*

Arnulfo Corte's cattle (right). *Clearing forests in the Panama Canal's watershed for cattle is causing the canal to silt up. Large dredgers struggle against erosion to keep the canal deep enough for ships* (below).

III

SRI LANKA: BALANCING THE ENVIRONMENTAL ACCOUNTS

'In a sense I have launched a war. I am not doing badly at all. I shall show others how to win this war.'
KIRANTHIDIYE PANNASEKERA THERA,
Buddhist monk

The Venerable Kiranthidiye Pannasekera Thera is the long name of a very short Buddhist monk, small and thin even compared with the slight build of most of his fellow Sri Lankans. When his face and body are in repose, Thera looks like a typical Buddhist monk of South Asia: the yellow robes of his order draped elegantly about his spare frame, his eyes dark and deep and his head shaved – a practice which has the effect of emphasising his prominent ears.

In many ways, this 33-year-old holy man also behaves like a monk. He lives on the simple food, usually rice and vegetable curry, left at his lodgings by villagers, and eats only before noon, nothing from then until the next morning. He is dignified and soft-spoken; he is calm, his gestures serene, and he never laughs in public.

'The Monk'

But the book he carries everywhere with him is not a Buddhist text; it is his diary and appointments book. He is rarely to be seen in repose, sitting and meditating. His twenty-three years as a monk have been dominated by activity, drive and a deep desire to get things done, to make things better – materially, as well as spiritually, better. Even his early calling to the priesthood was marked by trauma and emotion. A head priest had visited his parents' home and heard a boy – one of ten children – crying. On enquiry, the priest found that the boy was crying because he wanted to become a monk but his parents would not let him. The priest managed to convince the mother and father to let their son become a novice and live at the temple. Thera left his family; he was only ten years old.

After his period as a novice, he moved under the guidance of his superiors from place to place around Sri Lanka: a few years of study in the capital, Colombo, were offset by years spent in remote rural areas teaching children the scriptures. But in most places he insisted on doing things which monks do not normally do. He worked in the rice fields alongside the villagers 'because I was young, and I enjoyed the work'. He studied the local birdlife, and came to know the names and habits of the various species. While in one village, he found that his parishioners had difficulty in getting to the temple, so he organised the building of a road and helped with the work himself. Later, when he was in his mid-twenties, he became involved in a local political election, boosting the chances of one candidate against another, an activity strictly forbidden to men of his calling.

So in 1982, when the headman of the very remote village of Galahitiya in south-west Sri Lanka asked Thera's superiors to send a monk to the village to build a temple and guide the religious life of the villagers, Thera was chosen. It may well have been that Thera's elders felt that here at last was a straightforward job for this somewhat troublesome cleric, far off the main roads, where he would settle down and avoid controversy and where his eccentricities would go unnoticed. (To the villagers, Thera is known simply as 'the Monk', a practice that shall be followed here.)

When the Monk arrived in Galahitiya he saw immediately that his new parishioners needed material help more than they needed, or would be able to appreciate, an elaborate temple with a library and teaching facilities. Rather than building a *pirivena* temple as he had been instructed, he constructed a simple *pansala*: no more than a monk's lodgings big enough for himself and a picture of the Buddha. It was not the best way to start on the right footing with his new parishioners, or to stay on the right side of his superiors.

Then he carried out a survey of the village. Galahitiya is not a tight cluster of homes huddled closely to one another. Houses and families are dispersed throughout the valleys on the edge of the rainy highlands south-east of Colombo. A visitor can stand in the middle of Galahitiya without realising that he is anywhere near a village. The Monk found about one hundred families, none of which owned any land. All were squatting on government forest land, and had burned off most of the wild trees in order to farm it. The average income per person was the equivalent of £1.85 ($2.77) per month. Most of the houses were made from wattle and daub, with mud floors and thatched or tin roofs; there were no wells, and villagers fetched their water from the springs and rivers which bordered the village on three sides; only twelve families bothered to boil this water. Three couples alone practised any sort of family planning. There

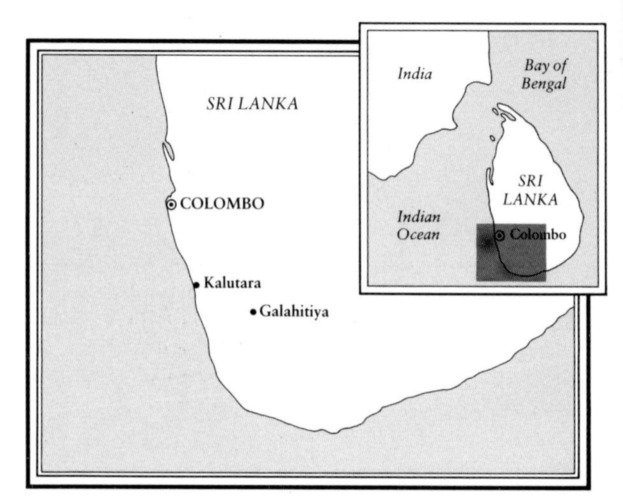

was no road to Galahitiya; to reach the nearest hospital it was necessary to take a boat across a river, followed by a two- or three-mile walk and a fifteen-mile bus ride. (The Monk's walk to the hospital while suffering from a bad toothache had made him feel this hardship personally.) Only two families drank milk with any regularity; only two owned lavatories; flies, lice and mosquitoes ruled the village. The villagers were not starving, but they were on the borderline; and floods and droughts often upset the fine balance between bare survival and starvation. About one-quarter of the families depended on government food assistance all the year round. The only positive statistic the Monk found in his initial survey was that '99 per cent of the families said they were willing to work to better the village if help were given'.

This dire catalogue of Third World poverty, malnutrition and isolation demands an explanation, for it is not typical of Sri Lanka as a whole. In terms of Gross National Product per person, the nation is among the world's 'Low Income Countries'; but in rankings of 'Physical Quality of Life' – an index which takes account of such things as infant mortality rates, life expectancy and literacy – Sri Lanka is rated higher than so-called Upper Middle Income countries such as Brazil and Malaysia, and is more than twice as well off as countries such as India, Pakistan and Bangladesh.

The Sinhalese kingdoms of ancient Sri Lanka (the island was called 'Lanka' before the British dubbed it 'Ceylon') had always seemed to their neighbours to be blessed with wealth and order. Traditionally, they exported rice to the rest of Asia. As long ago as the fifth century, King Dhatusena built a three-and-a-half-mile-long dam, forming a reservoir to provide water for his capital of Anuradhapura in the central northern plains. Sri Lanka is divided into a Dry Zone, roughly the northern and eastern sectors which make up three-quarters of the island, where the annual rainfall is seventy-eight inches or less; and a Wet Zone, the south-western quarter which includes the highlands, with over seventy-eight inches of rainfall per annum. During the first millennium AD, the island kingdoms of the Dry Zone ('dry' only in comparison with the wetter south-west) were dotted with elaborate irrigation works, reservoirs and gently sloping canals which made possible two rice harvests per year. The inhabitants lived close to the Buddhist ideal of harmony with their surroundings: the soil and water providing for the people, and the people nurturing these resources. These traditions are often recalled by modern Sri Lankan educators, agriculturalists and politicians.

Invasions from the north, warfare, population growth and finally the colonialism of the Portuguese, Dutch and British all combined to upset this balance. In the seventeenth century the Dutch introduced the alien concept of plantation agriculture, converting some Sinhalese officials into rich owners of estates growing cinnamon, pepper, coffee, sugar cane and tobacco. The British, who in 1818 unified the island under one rule for the first time in at least 1000 years, continued and modernised the plantation-building of the Dutch, concentrating first on coffee and then later in the nineteenth century on tea and rubber. The tea and rubber were grown in the wetter – and in the case of tea the higher – regions of the south-west, the nation's most fertile lands. The British actively discouraged rice cultivation and destroyed some rice irrigation works in order to encourage labour on the plantations. This emphasis on plantations and on crops for Europe meant that little attention was given to the complex systems of traditional paddy fields and gardens of spices, fruits and vegetables. Village agriculture declined as Sri Lankans, together with a great many imported Tamil labourers from southern India, went to work for cash. The profits went to Britain.

Since Independence in 1948, the government has taken measures to spread the profits of tea and rubber more equitably among the Sri Lankan people. But the continued stress laid on plantations has meant that this rich agricultural nation has been importing rice and wheat to feed its people ever since Independence. Only recently, with new government emphasis on rice cultivation, has the nation begun to move toward grain self-sufficiency. From 1948 until the late 1970s, rice production increased steadily, but so did population growth. Sri Lanka had almost 7 million inhabitants in 1948; over 16 million in 1985. Despite the spread of family planning, the population continues to double in number every thirty-two years. Although it has the second slowest doubling rate in South Asia, after India at thirty-six years, Sri Lanka's population is still increasing far too quickly for an island nation.

The Monk's village of Galahitiya lies in the heart of the Wet Zone, but on relatively poor land on the edge of mainstream tea and rubber country. Its 'squatters' are those who have been squeezed off other, overcrowded lands; some may have part-time jobs on nearby tea or rubber plantations. The 'squatters' cut down and burned trees to clear land for crops. Throughout Sri Lanka, but mainly in the Dry Zone where there is an obvious dry season, rice farmers have traditionally burned hillside forests during the dry months and planted crops as an insurance against rice failures. They sowed maize, a local grain (*Eleusine coracana*), cassava, chillies, bananas and beans. The cleared ground was used for such crops for a year or two, after which time forest was allowed to regrow for fifteen or twenty years. As only small plots were being cleared for this method of cultivation, called *chena*, entire hillsides were never left bare to erode away in the rains.

In Galahitiya, the Monk found families practising *chena* without leaving the forests time to recuperate. 'From my *pansala* I could see the forests being burnt for *chena*. And I felt pain in my heart. With every blaze that lit the sky I could feel the destruction being caused. For me the destruction of trees and plant life amounts to the destruction of life itself. I knew the destruction of trees and forests was an important factor in causing floods.'

However, the deforestation in the area around

Galahitiya was not only the fault of the villagers. In fact, the Monk blames 'racketeers' – illegal loggers – and legally contracted commercial loggers for 80 per cent of the destruction, the villagers for 15 per cent and cutting by the government Forestry Department for the remaining 5 per cent. In 1981, government foresters cut down valuable trees such as coconut and jack fruit in order to plant eighty acres of pine. In the process, they destroyed fifteen homes belonging to squatters. Villagers promptly uprooted almost all of the pines, but the incident reminded them of the fragility of their hold on the land. It gave them little incentive to protect and improve that land.

The destruction of hillside forests in a wet region can be disastrous: if the destruction is wrought by those who inhabit the area, it is almost a form of suicide. The Monk was deeply moved by a landslide which took place some years ago in a nearby valley. Soil and stones had tumbled down a cleared slope, killing forty people in the process. An entire family – children, parents and grandparents – was found dead, clutching one another in a final embrace under the rubble. When there are no trees to protect the soils from Sri Lanka's tropical monsoons and no root systems to guide rainfall into the ground whence it can replenish springs and streams, the 'watershed' sheds its water in destructive floods through the villages in the valleys. The rainwater, here as in the Panama Canal region (see Chapter Two), carries soil with it and causes the siltation of local rivers, which thus become ever shallower and more prone to overflow their banks.

'Floods have always been the main destroyer of villagers' livelihoods,' said the Monk. 'During floods there is no transport for days; the roads are submerged; the children cannot go to school; rubber tapping stops completely; rice fields are under water; the villagers become refugees and have to depend on the Red Cross, other relief organisations and the government. It is not only Galahitiya, but all the villages in the entire area.'

No, it is not only Galahitiya. It is a national problem. Some 100,000 acres of forest are lost every year, and only 20,000 acres are reforested; so the total area of forest lost yearly is larger than the island of Malta. In the early 1980s, Sri Lanka's hydroelectric reservoirs were so depleted by drought and so silted up by erosion from deforested watersheds that there were power cuts in Colombo.

Deforestation in the Himalayas seems to be the main cause of more and more flood damage in India and Bangladesh every year. Yet the fact that the rains sweep off the mountains into the rivers rather than filtering into the ground means that northern India's water table is drying up in some areas. A government survey in 1984 found that 2300 out of 2700 projects to install village water pumps in the northern hills of Uttar Pradesh state had failed because the water table had dropped. This led Dr Nalni Jayal of India's Planning Commission to predict for Asia a 'critical ecological crisis' on the scale of the Ethiopian famine in the near future. The Andes Mountains of South America suffer similarly: deforestation is followed by more extreme cycles of flood and drought; the rivers are loaded with sediment; their flow is erratic and wildlife is vanishing. Even in the drylands, deforestation and overuse of the land are causing radical swings from drought to flood and back. Some 40 per cent of Ethiopia was once forested; now the figure is estimated at 2 per cent. The rains that do come in the drought-prone highlands sweep down to either the Red Sea or the Nile, carrying millions of tons of topsoil with them each year.

The poor people on these fragile lands are both the partial cause and the prime victims of this overdrawing of environmental accounts. They must take from the soil and the forests in order to live, but their taking does not improve their lives; in areas such as the Ethiopian highlands it may not even keep them alive. They have nothing to reinvest in their environment. In the words of the US environmentalist Erik Eckholm: 'The stone wall of inopportunity facing the poorest billion people ensures continuing environmental degradation in large parts of the world.'

The problem is not technical. As the late British environmental writer Barbara Ward noted: 'The problems are rooted in the society and the economy – and in the end in the political structure. Foresters know how to plant trees, but not how to devise methods whereby villagers in India, in the Andes or the Sahel can manage a plantation for themselves. The solutions to such problems are increasingly seen to involve reforms in land tenure and economic strategy, and the involvement of communities in shaping their own lives.'

The village of Galahitiya holds about 1000 of the world's poorest billion people (the population has grown since the Monk carried out his original survey). The problems for Galahitiya are not technical, neither are the solutions to these problems. A Buddhist monk educated in the scriptures rather than economics had little difficulty in pinpointing the economic and political realities which were keeping his villagers poor: 'Most of the settlers in this village are destitute,' said the Monk. 'So they are compelled to go out and work in estates or rich households. However, this income is not sufficient to help them make "their" land fertile. In brief, they are not financially strong enough to prepare their land for effective cultivation. Of course there are various government schemes to help farmers financially to cultivate their land. But the problem in this village is the legal title. Government help only goes to those whose land ownership has been properly recognised by the appropriate government departments.'

Before he could organise his villagers to improve their plots, the Monk had to secure for them land which was legally theirs, which they would see as being worth the work of improvement. It was no good to invest hard work in land on which the Forestry Department might at any moment plant pines. Securing titles to the land was theoretically possible, because the government operates various programmes as a means of providing land for the landless. In reality, however, the process is a nightmare of applications, appeals and procedures which can take years. The possibility of success was seen by the illiterate people in an isolated village as a miracle beyond their powers. The 78-year-old village headman, Kudhemis Fernando, who had originally sought a monk to build a temple, began the process before the Monk arrived in the village and had reached the stage of receiving an 'authorising paper'. But when the local Land Registry Office demanded to see other papers of which he had never heard he gave up. Fernando's experience did not inspire other villagers, who did not see themselves succeeding where their headman had failed.

The Monk took the matter into his own hands. He wrote letters, asked the villagers to sign them, and then followed up the written applications by presenting himself at the office of the District Land Officer, Wijesena Sinhabahu, in the district capital of Kalutara on the coast.

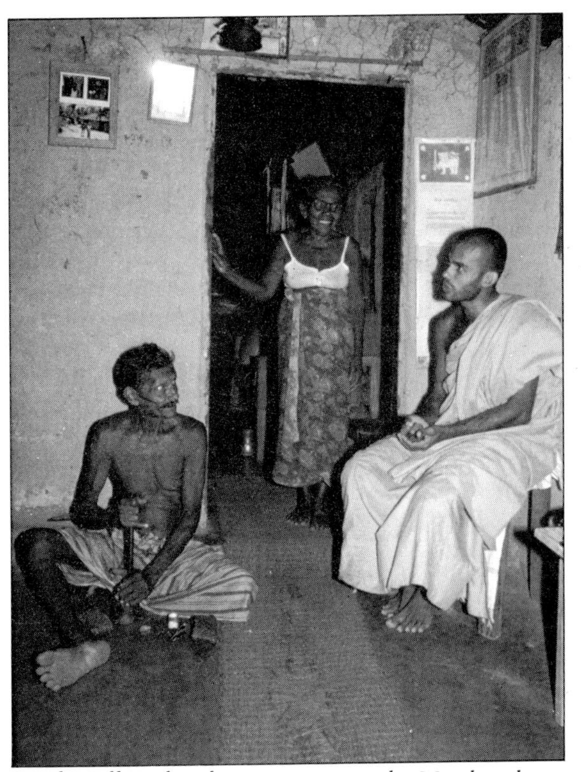

The village headman entertains the Monk, who keeps a close watch on village politics

This confrontation between Buddhist monk and bureaucracy has been a complex and controversial affair. Sri Lankans are not used to refusing a monk's requests, both out of respect for religion and because monks rarely ask for much. But this monk strode with Kudhemis Fernando into District Land Officer Sinhabahu's office and demanded land for over 250 families. The crowd of people in the packed room, all there on similar errands, melted away. A chair was found for the Monk; he was listened to with grudging respect and addressed as 'Reverend Sir'. Lining the room were ranks of overflowing filing cabinets out of which protruded the dry edges of applications yellow with age.

The Monk hammered away quietly at the bureaucrat, piling up evidence of both need and eligibility. He reminded Sinhabahu that he had the backing of the local Member of Parliament and that the manager of the Agricultural Development Authority headed a committee working on the development of Galahitiya. 'Accordingly, I would like to remind you, Sir, that unlike other work being done in the district, our Galahitiya work should get special attention.' But, when Kudhemis Fernando joined the conversation to back up the Monk's arguments, the District Land Officer demonstrated his usual treatment of supplicants who were not holy men, even though they might rank high in their village hierarchy; he brusquely accused the headman of trying to steal land for himself.

Finally, however, Sinhabahu capitulated and promised quick action, a promise he kept. In early 1985, 130 families – including that of Kudhemis Fernando – received legal title to an average of two acres each, and in the summer of 1986, a further 115 families received similar titles. The Monk had worked the miracle. He had not only won land; he had won his flock's respect; he had succeeded where the headman had failed. Previously, some villagers had regarded their monk with a good deal of suspicion and dissatisfaction. They had wanted a spiritual leader who would build a temple and then stay in it, meditating, organising religious functions, and seeing to the rites of passage of his parishioners: births, marriages and deaths. Instead they had been given an eccentric who told them how to farm their land, build their houses, boil their water – none of which was really a monk's business at all. Now that he had gained for them the treasure of land, they were willing to begin listening to him when he told them that they must stop the practice of burning trees, which had been going on for thousands of years.

In 1983, the Monk had written a sophisticated 'Project Proposal for the Development of Galahitiya' which contained the results of his early survey, and a suggested budget of £207,500 ($311,250) which would cover everything from building a road to the village to helping villagers plant tea and erect new houses. The proposal was aimed at all possible donors, from the local Lions Club to the government of Sri Lanka and any foreign organisation which might listen.

The Monk's Development Project has not attracted any government funding; some said it was because of the Monk's earlier political activities. But, even without funding, the villagers built the dirt road he said they needed. The road was necessary not just in order to transport villagers out of Galahitiya to schools and hospitals but to bring surveyors, agricultural experts and even Members of Parliament in. Before the building of the road, only the lowliest government officials had visited; but the local MP came in person to hand out the land titles, once he knew he could be driven to the village.

The Monk, whose knowledge of local bureaucracy rivals that of most local government officials, brought representatives from the Rubber Control Department and the Small Export Crops Board to the village and won a promise of the supply of free rubber trees to be planted in late 1986. He encouraged each villager to devote an acre of land to this cash crop, a further half an acre to trees as a means of protecting the soil and half an acre to the house and the 'home garden'. The home garden, which looks to the untrained eye like a jungle, is a tiered allotment which is traditional throughout Sri Lanka and Indonesia. Carefully ordered layers that allow the right amount of light through to the various plants mean that coconut palms, banana, mango and kapok trees, and jack trees with pepper vines climbing them, may be grown together. At lower levels grow vegetables, chillies, yams, onions, pineapple and perhaps a few tea and coffee bushes.

Turning to the question of housing, the Monk discovered that the United Nations had designated 1987 as 'International Year of Shelter for the Homeless' and that Sri Lankan Prime Minister Ranasinghe Premadasa had taken a personal interest in this effort and established his own national 'model villages programme'. The Monk met his MP and requested that the district be included in the villages programme so that people could build their own sound brick houses which would be able to withstand floods; the MP promised to do this.

The Monk, given his obvious flair for politics, was asked why he did not abandon his ordination and become a politician. He replied, 'As a monk, I have influence from the day I'm ordained until the day I die. A politician only has influence for the seven or so years he is in power. Also, a monk can help everybody and get involved in everything irrespective of party politics.' And get involved he does, daily, in the lives of all. He strides through his parish carrying his diary and one of his only two other private possessions: a heart-shaped fan for the hot season, and for the rain a black umbrella with which he points to various parts of the landscape during his frequent lectures on agricultural development.

He visited Chalosingho, a poor *chena* farmer who lives in a one-room mud hut with his wife and four children. After helping Chalosingho to write a letter to the land officer requesting land, the Monk knelt to pick up some grains of the sandy, infertile soil of the hillside plot and let it run through his fingers.

'Was it you who first started cutting the forests down in this area, during the days of the thick forests?' the Monk asked quietly. The answer was affirmative. 'There must have been beautiful streams and brooks around here at that time?' Again the answer was yes. 'None of that is left now, is it? That is one of the damages of *chena* cultivation. Streams, brooks, springs all gone dry. Well, whatever is said and done, let us forget our past mistakes and plan for the future. Writing letters to get you land will not be enough. You must regularly visit the *pansala* and work with our associations in the village. Only then can we keep up the pressure through these associations with the officials and get the ministers to bow to our demands.' One association which the Monk established in his village was the 'Friends of Nature', designed to educate the farmers and their children in the management of environmental resources.

Other visits on the Monk's rounds are more optimistic. Gunapala is a part-time driver who earns some £17 ($25) per month on a nearby tea plantation. But he is an ambitious man, and seeks his fortune in

The Monk and Gunapala in his well-ordered, irrigated garden – a model for the village

his newly acquired land. He secured his two acres from the Land Office, and then quietly took over two more. All of the village, the Monk included, turned a blind eye because of the wonders Gunapala worked on the land: irrigation ditches, drainage ditches, walls to stop erosion and put run-off water back into the soil. He grows enough rice to feed himself, his wife and three children. His other crops are cloves, pepper, pineapples, mangoes, rubber and some medicinal herbs. He had just planted 100 coconut trees.

'Well,' said the Monk, 'if you've got 100 coconut trees, you are now an estate owner. The land of an estate owner should look attractive to our villagers, so they will want plantations like yours on their lands. You must take off your dead branches and spread them around to form a new layer of soil here. Once that happens there is no need to import fertiliser from abroad. Your land will be our textbook. Once our people have learned from your work the whole village will change. You won't find a village like it anywhere in Sri Lanka.'

Much of what the Monk does, and has others do, is for the sake of example. Each day he works in his own tree nursery, a project started to convince villagers that they need not buy seedlings from outside but could produce their own. The Monk's nursery has provided about 6000 seedlings, with which five acres of hilltop have been reforested. Caring for the nursery is perfect work for children, whom he sees as the hope for carrying on his efforts: 'I particularly make sure I get the children involved in plant nursery work. By doing this I train the young generation to do the exact opposite of what their fathers had been doing. The older generation destroyed the forests; now the children are going to recreate those forests by planting trees exactly on the spots their fathers cut them down.'

When dealing with the government and in his own formal development project proposal, the Monk emphasises first that if his goals can be realised the people will no longer be living largely on government assistance but will be contributing to the national economy. His second key point is that, although it is small in scale, his work can stretch far: 'In a sense I have launched a war. I am not doing badly at all. I shall show others how to win this war. Let me tell you something: the *pansala* and the land around it are an example for the villagers. And I am constructing Galahitiya as an example to the whole country; and if we can make a few hundred people work like we do, then we can make Sri Lanka an example for the whole world!'

The Monk has a friend and adviser in the form of Dr Sarath Kotagama, a lecturer in the Department of Zoology at Colombo University. His is a young environmental activist, with hair worn parted in the middle and hanging far below his shoulders, an effect which together with his thick beard gives him a disturbing resemblance to paintings of Jesus. Dr Kotagama, who visits Galahitiya often and frequently receives the Monk at his home in Colombo, describes the Monk's work in terms of development jargon, but remains enthusiastic about it: 'What the Monk is doing in Galahitiya would be called a "small-scale community development programme". Many people see such programmes as ineffective and unimportant. But the fact is that they are the ones which will be successful in the end. Most programmes, imposed from above, do not succeed because they are too uniform. But a given problem is never uniform because the people are different, the environment is different. The Monk has realised this, and turned the people of Galahitiya into a group and given them a programme which will be sustainable,' said Dr Kotagama.

'In Sri Lanka in most temples we have what we call a twelve-month lantern. It is lit at the beginning of the year and is supposed to be kept burning all year by the people who come and put oil in the lantern. The Monk has done something like this in Galahitiya. He has lit the lantern for people's development, and success will depend on how the people continue to put oil in the lantern.'

The French have the colourful word *animateur* to describe a person who instigates a local project and keeps it going until it develops a momentum and life of its own. It is essential that the local people do most of the work, and that they feel the needs which the 'community development programme' is designed to address, but in most cases an outsider is needed to come in and stir things up, to provide the resources the people might not have organised by themselves (in the case of Galahitiya, land rights and rubber trees). Local initiative by itself is rarely

The Monk and the headman of the village of Galahitiya, Sri Lanka. They have been in the city seeking land titles for the villagers.

The Monk prays for strength in his 'war' against environmental destruction (above). *Village children replant a hillside, repairing the destruction of their fathers* (left).

Burning off the hillsides to make dry-season gardens (opposite above). *Forests are also cut down for fuel* (opposite below) *and are not given time to grow back.*

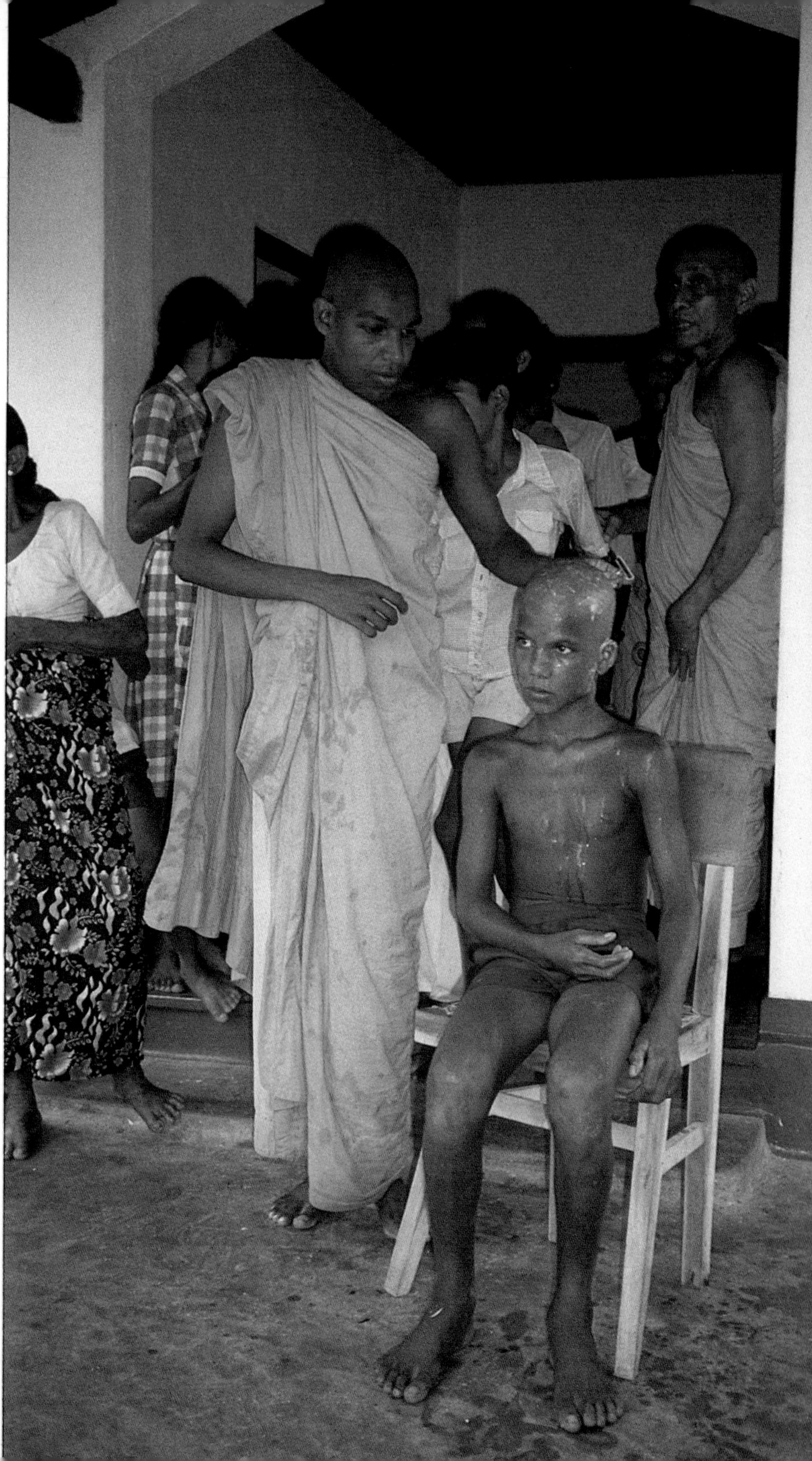

The Monk shaves the head of fifteen-year-old Saliya, and the boy then becomes a monk to carry on his elder's environmental and spiritual work

The village of Korr, Northern Kenya (left). *Thorn livestock corrals surround the mud buildings. D'igir* (below) *takes tea from his wife in their camp on the dry plains.*

D'igir's son with the herds. Modern school syllabuses do not teach nomads' children to herd in the desert. D'igir ensures his children learn the traditional skills.

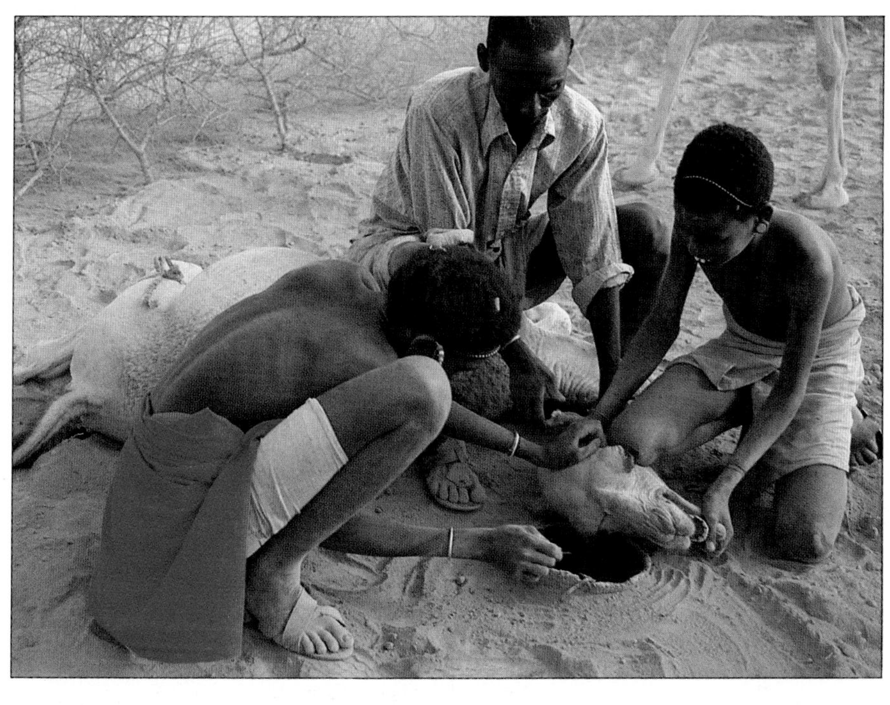

Bleeding a camel (left). *The traditional Rendille diet of milk and blood requires no cooking, thus no firewood. Herders gather at shallow, hand-dug wells in the bush* (below).

Women at the deep wells in Korr (right). *Herders gather at a UN-sponsored livestock auction in Korr* (below). *Such auctions cut out the middle-man, bringing bigger profits to the herders.*

The Monk, his botanist friend Dr Kotagama, and some helpers in the Monk's tree nursery

enough, and when national governments pretend that it is, they are usually shirking their own responsibility. In a study of projects to bring clean water to villages and shantytowns in Africa, Asia and Latin America, the Indian development writer Sumi Krishna found that each successful scheme had benefited from an *animateur*: a concerned district official, a returned university student, a foreign volunteer, a nurse, a priest. A monk provided the impetus in Galahitiya. If the *animateur* is of too high a rank or is in charge of too many projects at once, he or she will lose sight of the specific needs of the local people and environment which Dr Kotagama feels are so important.

But, if work such as the Monk's in Galahitiya is to spread, the government must be involved; it must want to undertake land reform, to shift resources to the poor so that they can invest in village building and thus in nation building. 'You cannot carry out such a programme on a national scale without government participation,' argued the Monk. 'Individuals cannot do it.'

Dr Kotagama was certain that, as long as the Monk was encouraging his villagers, the lantern he had lit would stay alight: 'He may look pious, serene and calm, but I can tell you he is not. He is very pushy and a very good politician. He always gets his way. He does not understand the word "no". He will come to your doorstep at perhaps 5 o'clock in the morning and get his needs attended to, and he continues to do this and convert others to his path.'

More important for the future of Sri Lanka, 'The Monk is not alone!' said Dr Kotagama. 'There are many other such people in Sri Lanka as well as elsewhere who are trying the same process. It's a spreading thing, you know. People are doing it everywhere.' It will take a nationwide effort to reach the roughly 6·5 million Sri Lankans who the Monk estimates are in more or less the same condition as the villagers he met on his first visit to Galahitiya.

A meeting in Colombo in late 1984 supported the view that the Monk represents part of a growing movement in grassroots development. The meeting brought together over 130 small local groups from all over the country, with the common concern of environmental issues, many representing only one village, such as the Monk's 'Friends of Nature' group from Galahitiya. The big national groups were also brought in, organisations such as Nation Builders, a well-disciplined collection of students who carry out development projects; the March for Conservation, headed by Dr Kotagama; and the Environment Foundation, a group of young lawyers and law students who are active in environmental law. Some of the older, more élitist conservation groups concerned mainly with saving wildlife were notable by their absence.

'It was an attempt to collect all the frogs who have been croaking everywhere,' said K. H. J. Wijedasa, secretary to the prime minister and chairman of the Central Environmental Authority. The meeting became a permanent body, the Sri Lanka Environment Congress, in which all non-governmental organisations were invited to take part. Representatives of governmental organisations can be members in their personal capacities but, in order to ensure the Congress' ability to criticise the government when and where necessary, only the non-governmental bodies have voting rights. Dr Kotagama helped to draft the organisation's constitution and became its 'co-ordinator', while Galahitiya's Monk, the Venerable Kiranthidiye Pannasekera Thera, was elected secretary.

'Here we have discovered a large number of village-based organisations whose existence we hardly knew of until our first meeting,' said Donatus de Silva, a Sri Lankan working with the London-based environment and development organisation Earthscan, which helped to structure the Congress. 'No country that I know of has this combination of membership in one body concerned with conservation of the environment.'

Thus, the Monk became an officer in a national body, but his real work remains in the village of Galahitiya. There, in the spring of 1986, he was relaxing after an extremely tiring day. It was Vesak, the celebration of the Buddha's birth, enlightenment and death all in one, the most important day in the Buddhist calendar. It is a day for lighting lanterns made from bamboo and paper, for dancing, for spiritual consecration – all of which activities had to be organised by the Monk. The Monk also used the occasion to continue his preaching on trees, as the Buddha was born, received enlightenment and died under a tree.

On the previous day he had officiated at another type of ceremony; he had gathered together fellow elder monks to help with the ordination ceremony of the fifteen-year-old boy Saliya. The Monk had been born and raised only fifteen miles from Galahitiya, in a bigger village with a grand temple, a true *pirivena*, a teaching monastery with many monks. He brought Saliya to Galahitiya from his own home village, and had been training him in the basics of Buddhism for some months. His idea in bringing in Saliya was both to have an understudy through whom to pass on his teachings and to give the villagers a more 'traditional' monk who would mind the shrine and be on hand for the ceremonies of births and deaths when the Monk was about his development work, perhaps in the capital. At the ordination ceremony, Saliya's head had been shaved; he had been given his yellow robe and begging bowl; and his mother had cried – all according to custom. Almost immediately the impish Saliya, once all smiles and noise, took on a quiet dignity, that of a monk.

The Vesak celebrations had been even more elaborate than usual because the Monk – with his characteristic flair for public relations – had taken some money from a visiting BBC television crew and gone to the next village in order to hire traditionally costumed professional dancers to augment the spectacle provided by the efforts of Galahitiya's enthusiastic amateurs.

The holy day, being a celebration of both birth and change, seemed to the Monk an appropriate time to look into the past and into the future, at what had been done and what remained to be done. He remained defensive, but not apologetic, about his as yet small contribution to the spiritual life of his community, and his much greater concern for its financial, material, agricultural and environmental

The Vesak celebrations in Galahitiya

development. He found his example in a story from the Buddhist scriptures: 'Once the Lord Buddha visited a wealthy merchant's household. In the same household was a poor man, hungry and exhausted, who had been out all day searching to no avail for his only ox, the source of all his livelihood. The Buddha was about to preach a sermon, when he noticed the poor visitor. The Buddha could tell that the man was ready to grasp his philosophy. "But," the Buddha thought, "this man is starving, and a starving man will not be able to follow my sermon." So he sent the hungry man his own plate, and only then did he preach.'

The Monk pointed to a poster entitled 'The Caretaker' which hung on the wall of the *pansala* near the picture of the Buddha. The poster related an incident which happened 2300 years ago, when a Sri Lankan king was hunting with a bow and arrows in the forest. An apparently poor man in a simple robe shouted at him to stop. Not used to being given orders, the king asked the stranger to explain himself. The plain man was the Buddhist monk Mahinda, son of India's Buddhist Emperor Aśoka. He described his religion to the king, emphasised its teachings on the interrelatedness of all things, and explained to the king his responsibilities for his forests and the living things in them. Thus, Buddhism was introduced to Sri Lanka with a lesson involving the management of environmental resources. The Monk uses the poster today to teach his villagers that it is the job of all people to husband these resources well and to see that their leaders do likewise.

The Monk estimated that – despite the secure deeds to land, the tree planting, the road building and the promised rubber trees – he had accomplished only 1 per cent of his goal. (On an earlier, less tiring day, he had put the estimate at 5 per cent.)

'In another twenty-five years this village will be richer. People will have money in their hands; plants and trees will grow on their lands; houses will have been built; children will be literate and educated. Only then will the people in Galahitiya have the strength to build a *pirivena*, a proper temple, like the beautiful temple where I was ordained as a child.

'But first the environment; then the temple will emerge from that.'

IV

KENYA: FROM SOIL TO SAND, AND BACK

'The pastoral life is very hard, but we like it. We depend on these animals. That is how we get food, how we get our living, how we get everything.'
D'IGIR TUROGA, Rendille herder

D'igir Turoga is probably in his late forties, but he is not certain. He knows that he started herding camels in 1951, and boys are not given this awesome responsibility in Rendille society until they are about twelve.

He had not seen anyone wearing 'clothes', in the European sense, until he began school at the age of thirteen, and the sight frightened him. Today, however, his own daily costume consists of a Western-style, lilac patterned shirt, worn with a faded *khanga* (sarong) of flowery cloth and sandals, the soles of which are made from car tyres. He keeps a pair of trousers hidden away for trips to Nairobi in the distant south. He also keeps his Christian name, James, in reserve for Nairobi and for dealing with visiting missionaries and Europeans.

D'igir Turoga's bearing is more striking than his costume. His nickname 'D'igir' means 'man of upright stance'. Indeed he strides forward across the plains of northern Kenya with chest out and shoulders back, hatless, revealing the narrow face, thin nose and high cheekbones typical of the dryland nomads who inhabit East Africa.

D'igir Turoga

In his lifetime he has herded in the desert; he has fought Somali raiders; he has lost cattle to lions; he has been a policeman, a deserter from the police force, a fugitive in the bush, a security guard in the big city of Nairobi, and he has converted souls to Catholicism as an 'assistant missionary' for a Catholic mission.

Today he has two lives, one traditional and one modern, as reflected in his two homes. He divides his time between the temporary hide and pole tent in his *fora* camp in the bush, where he keeps most of his herd, and a more substantial round hide dwelling in Korr, the dusty, tin-roofed town which sprang up in the early 1970s and which is now home for 2500 people. Korr has a deep well, several smaller wells, a clinic, a school, a Catholic mission, and a scattered collection of the typical Rendille huts. D'igir keeps some camels and goats in Korr, and the three eldest of his five children go to school there, living with their grandmother. In Korr he also conducts his business as chairman of the Pastoralists' Association, an organisation devoted to improving the care and marketing of livestock.

Despite his schooling and experience, he spends much of his life in movable camps on the roadless plains at least a day's walk from Korr. He herds sheep, goats and camels over dry, stony ground – as the Rendille people have done for centuries. An outsider, even a fellow Kenyan, might pity him, his wife and children, seeing them as victims trapped in the old ways as their country has moved boldly ahead into the twentieth century. But that is not the way D'igir sees himself. He believes that he can combine the best of the old ways with the best of the new, and become prosperous in doing so. He is bold enough to think that his whole tribe can do so also.

Time has moved slowly here, on the vast, flat drylands of northern Kenya, in the southern reaches of the Koroli Desert with Mount Marsabit to the north-east and the southern tip of Lake Turkana to the north-west. The fossilised remains of humanity's remotest ancestors have been found on the shores of that very lake. Those early humans of *Homo erectus* stock who walked these plains $1\frac{1}{2}$ million years ago may have lived in huts not unlike the temporary structures D'igir's wife puts up as they move with the livestock.

Given the slow pace of change here over millennia, the spontaneous settling in towns of most of the Rendille tribe over a part of one man's lifetime has taken place with the blink-of-an-eye suddenness of a lightning bolt. When D'igir was young, the Rendille were always on the move, driving their herds great distances. He recalled: 'Before, if it rained even beyond Marsabit [fifty miles away] people would move with their livestock and follow where the rains were. But now they're settled in one place, and they are not moving.'

A 'settled Rendille' is somewhat of a contradiction in terms; it is like saying a 'shy politician'. The only way to live off this land is to move across it. Rendille land is flat, rocky, dotted with small trees and thorn bushes. There are hills on the eastern and western edges of the plains, and they leap out of the landscape so suddenly and starkly that they look like a child's attempts to draw mountains.

On the government map of Kenya, the realities of

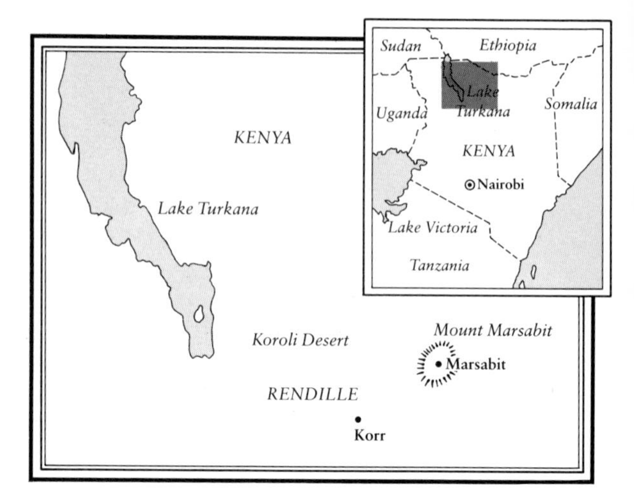

D'igir's land are summarised in a few words. 'Koroli Desert', it says, and just above this is the warning 'Liable to Flood'. There are two rainy seasons here: the long rains in April and May and the short rains in November. But the word 'season' suggests that the rains are much more predictable and steady than they are in reality. Add together rainfall from the long and the short rains and you arrive at only six inches on the Rendille's central plains in an average year. But the word 'average' means nothing here, because 'normal variation' from that average can bring only one and a half inches of rain one year and eighteen the next. Variation from place to place is even more erratic than variation from year to year. Rains can be heavy when they do come, and water often rushes off the baked ground in flash floods; thus the apparent contradiction of a flood-prone desert.

It may suddenly rain in a valley for the first time in ten years; and it may not rain there for another decade. Therefore, the Rendille did not so much follow the rains as chase them, rushing to get their animals on to new grasses, which are more easily digested and converted into milk than are the drier, older shoots. From the herders' viewpoint, their animals are miracles of efficiency: able not only to process grass, useless to humans, into milk, blood and meat, but to cover great distances in doing so. A camel is a highly mobile, low-maintenance protein factory. The herders' ability to chase the rains was limited mainly by the availability of water: one cannot move on to fresh grasslands if there is no water, either in wells or on the surface, for people and animals to drink.

Looking over the baked plains from Korr, it is easy to see why the traditional Rendille herders planted no crops; they plant few today, except in small, watered plots near the town. Not surprisingly, livestock long ago became the cement of Rendille society. It is not just that men 'bought' brides with camels. The exchange of animals linked the families in the clans and linked the clans together in bonds of obligation so that a man always owed livestock and was always owed. A herder could expect help in crises, such as drought or when disease struck humans or animals, because there was usually a herder somewhere, unaffected by the crisis, who owed him animals. The Rendille were really not so much a group of herders as a complex corporate structure. The clans resembled linked subsidiary companies, with inter-clan marriages ensuring that shareholders had 'stock' in many different clans. Traditionally, livestock were not sold for money; they were themselves money, providing far stronger connections than money could.

The old ways are not dead, but they are changing quickly. Though there are few Rendille in Rendille country, there are many more than there used to be; at the latest count a little less than 20,000.

'Medical care has helped the population explosion,' D'igir maintained. In 1884, an outbreak of smallpox killed virtually an entire clan, and the whole age-set was named after the only surviving *moran*, or unmarried warrior. The age-set system, common to many of Africa's herding tribes, links all young men born within a given span of years into a fighting fraternity. Following initiation into an age-set, a young man in the old days might serve fifteen years of warrior status – taking camels to distant pastures, guarding herds and raiding the livestock of nearby tribes. A *moran* is not allowed to marry or have children, so the custom kept population growth low. But as the young men's herding duties have diminished, and as the Kenyan government has laboured by means of education and police patrols to stop stock raiding, men spend less time as childless warriors. They marry younger and have larger families than their predecessors.

D'igir Turoga, always a pioneer and innovator, was one of the first to settle in Korr. After a two-year primary education, he had as a teenager become an assistant policeman. But he felt that the job was beneath him, and one day discarded his policeman's uniform, took up the two spears of the Rendille *moran* and deserted, herding his camels into the desert where the authorities were unlikely to find him, or even to look. While herding and trading, he ran across Catholic missionaries and began to read their books. D'igir was persuaded to put aside his spears, to wash the red ochre from his body, abandon his warrior ways and teach Christianity in villages. He travelled from clan to clan, handing out free tobacco in order to guarantee his welcome and to secure the people's attention, before delivering his teachings from the Bible.

In 1972, D'igir met the Italian missionary Father Redento and joined him in the small settlement that was later to become the town of Korr. There were three wells and a broken-down borehole from the 1950s, but these provided enough water for only thirty cattle after the humans' needs had been met. Redento went to Italy where he raised £200,000 ($300,000) and on his return paid for the digging of twelve wells and a new borehole.

The proliferation of deep, permanent wells dug by outsiders in Rendille country has had even more drastic effects on the tribal life-style than has population growth. These wells, and recent droughts, have encouraged the Rendille into permanent settlement and into idleness. When drought struck the Marsabit district in 1971, about 20 per cent of the region's cattle and high proportions of other livestock were killed. Rendille livestock expend most of their energy simply in surviving and do not give very much milk, so one family needs at least four milk camels and 120 sheep and goats in order to meet its protein needs. Three big pack camels are required just to transport a household and the poles for the family hut. Thus, a loss of even a relatively small percentage of stock meant the difference between survival and starvation for many families in the 1971 drought, and those who lost their pack camels could not travel.

Wells are not a new concept to the Rendille. Scientists studying the parched district recently found that water could be obtained almost anywhere in the area

The Catholic church in Korr. Missionaries built up the town, and their wells helped to settle the nomads.

by digging a shallow well. The Rendille had been digging such wells by hand for centuries. It is hard work; water is handed up from person to person in buckets or bags (and a camel which has gone without for fifteen days needs a large quantity of water). Unless such wells are shored up, their sides soon fall in. Also, as the wells reach down only to the water, it is soon used up. Obtaining more would mean digging deeper. For the Rendille, however, the transitory nature of the wells fitted perfectly with the transitory nature of the herders' survival strategy, and both factors prevented grasslands from being over-used.

Father Redento established a clinic in Korr; shops sprang up; a school was built; clothes and food were handed out. Many Rendille ate grain for the first time in their lives – a radical change from the traditional diet of milk and blood drained from the neck of a living camel. (Grain, unlike milk and blood, requires cooking, so the greater the amount of grain handed out, the greater the number of trees felled around the town to provide firewood.) Thus the herding families, having just been savaged by drought, found in Korr an almost magical source of permanent water, free grain, modern clothes and medical help. They settled in the town and near by, and the very real desire on the part of the missionaries to be of help turned the herders into beggars.

'My own clan – my *manyatta* [the large mobile clan camps] – had 120 huts, about 750 people. It was very big. Now the clan is broken up in various places,' said D'igir Turoga. In fact, three-quarters of the entire Rendille tribe have settled either in Korr or in the larger town of Kargi to the north. 'A long time ago, when the Rendille were nomads, they were moving from place to place, they used to help poor families. They used to adopt them. The [poor family's] children would look after their livestock, and the head of the family who adopted the other family would support them by giving them milk animals and portions of meat,' remembered D'igir. 'But now with the famine relief, a lot of people are used to that. Now people are accustomed to the famine relief, and the problem is that they are just taking.'

Did Father Redento mean to encourage the Rendille to settle, to destroy the herders' ability to make their own way in the world? 'In the beginning,' said D'igir, 'Redento encouraged pastoralism. But it was discovered that, when they got water, people wanted to stay. A lot of old people in the *manyattas* settled here; they stayed here while others took the animals around.'

However, the animals are not taken far. Cattle need water every second day; they must therefore graze only about ten miles from the nearest source. Sheep and goats can go for four days without water, so they can use pastures twenty miles from the watering point. D'igir Turoga must bring his own sheep, goats and camels every few days from his *fora* camp, where there is no water, to the Korr borehole. Traditionally, the Rendille depended almost exclusively on camels, these animals being best adapted to the desert; but the increased availability of water has encouraged the keeping of more goats, sheep and even cattle. Goats are particularly voracious browsers, eating young seedlings before they can grow into trees, and their hooves break up the soil, leaving it exposed to wind erosion. Cattle are particularly vulnerable to drought.

The area around Korr was soon overgrazed. Less rainfall trickled underground through root systems; more ran off the soil in the form of floods. Wind took away exposed topsoil and left sand behind. Before long, there was a circular desert around Korr within a radius of twenty miles – and all because of the wells. 'Man-made desertification' is what geographers call this syndrome; 'hard times' is what the Rendille in Korr call it.

Desertification, the process whereby humans ruin productive drylands, is a worldwide disaster. When tens, perhaps hundreds, of thousands of people died of famine in the 1970s in the Sahel, the strip of land across much of Africa just south of the Sahara Desert, the United Nations began to organise scientists to study the syndrome. Experts found that 12 million square miles of land – one-fifth of the planet's land surface; an area as large as the USSR and China combined – were threatened by desertification, together with the 80 million people who live on that land. That is to say, D'igir Turoga, with all his experience, labour and aspirations, multiplied by 80 million. The UN found that every year 80,000 square miles of productive dryland, an area larger than the nation of Senegal, deteriorates until it has no economic worth.

Desertification is not just an African or a Third World phenomenon. Stock raising and the over-use of groundwater are threatening rangelands in the western USA. Much of the Australian outback is also in jeopardy. A long, broad ribbon of threatened land runs down almost all of the drier, western coast of South America. At least 40 per cent of Asia is at high risk, mainly in the drier areas of India, Pakistan and Afghanistan. Desert is also spreading rapidly in the Asian part of the USSR, as that nation seeks desperately to increase agricultural production. Overgrazing is not the only cause. Desertification may occur where the land is not given a chance to 'rest' between crops. In some areas too many trees are cut down, and the land exposed to erosion. Badly built and maintained irrigation systems also turn good land into wet, salty deserts, wasting not only the land but the large sums of money spent on the irrigation systems in the first place. Irrigation has waterlogged some 11,500 square miles of Indian farmland, and big canal irrigation projects cost £14 million ($21 million) per square mile. Bad irrigation can therefore be a very expensive form of desertification.

When the UN Environment Programme (UNEP), which gathered together much of the early scientific work on desertification in the 1970s, examined anti-desertification progress in 1984, it found almost complete failure around the world. 'Governments do not see desertification as a high-priority item,' wrote a UNEP expert. 'Lip service is paid to combating desertification but the political will is directed elsewhere.' As if to prove his point, famine swept the drylands of Africa again in 1984–5, with much more disastrous effects than in the early 1970s. Once more, people died because of desertification compounded by government lip service and lack of political will; 30 million lives were threatened and 10 million people fled their homes. No one knows how many died, either in the Sahel drought of the 1970s or the more widespread disaster of the 1980s. Desertification is a mass killer which has yet to find its rightful place on the international political agenda, as it affects only the poorest people on the poorest lands, far from the big cities.

Drought struck Korr and northern Kenya yet again in 1981, reinforcing the need for the grain handouts which had begun in 1971. They have not stopped since. Families have not moved with their herds for so long that many women of Turoga's age and younger have forgotten, or have never learned, the complex techniques of loading one house and household contents on three camels. Men have forgotten how to study landscape and vegetation to find the places where water lies just a few feet underground. Some Rendille have come to envy D'igir's continuing mobility: 'I remember some women talking, when I was at the wells with my wife. They said: "You people are much better off because you are going from place to place. Most of us have not even been out. We cannot even manage with loading up the camels."'

Between the two droughts in Norther Kenya there began another process which may eventually undo some of the damage caused by both the wells and the weather. In the mid-1970s, Rendille land was invaded by teams of scientists of various colours, nationalities and disciplines. These anthropologists, rangeland experts, soil scientists, hydrologists and meteorologists were known collectively as 'IPAL' – the Integrated Project in Arid Lands. IPAL had been set up by UNESCO, UNEP and the Kenyan government in order to find ways of improving the lot of the nomadic tribes of northern Kenya. (The IPAL project also covers the lands to the north, south and west of the Rendille, which belong to Gabra, Samburu and Turkana nomadic pastoralist tribes. It encompasses more than 8500 square miles, an area twice the size of the island of Jamaica.)

There are many reasons for the failure of aid projects which are designed to 'develop' traditional cultures; some have even done more harm than good. All too often they break down because they are dreamed up in places such as London and Washington and then applied in the developing world by foreign teams with little idea of the life-styles of the people to be helped. Development consultants, who are based in European and North American capitals but work in the Third World on two- to three-year contracts, may have big budgets, but they have little time. They can rarely take years to study the peculiarities of the local people, their ambitions, their skills, their own ways of coping. Often there is only enough time to begin the building of a dam or an irrigation system, or the introduction of a new crop. But 'improvements' which may have worked well enough in the drylands of California, Australia or Israel do not necessarily succeed elsewhere.

The IPAL experts tried to avoid this classic error by taking their time. Rather than immediately recommending new techniques and building new things, they spent time studying the region closely. They decided early on that the most important variable in the region was not the rainfall, water tables or the soil, but the people.

Dr Walter Lusigi, a Kenyan range ecologist who puts his faith in the herders

Dr Walter Lusigi, a Kenyan range ecologist and former project manager of IPAL, places humans firmly at the centre of the problem, maintaining: 'Man can justifiably be regarded as the dominant element in the grazing land ecosystems by virtue of his overwhelming impact upon them, exerted largely through his domestic animals.'

Dr Lusigi is a large, precise man who was trained in West Germany. He is as comfortable in khaki safari clothes in the African bush as he is in a Western suit on the international academic conference circuit. He has devoted many of his lectures at such gatherings to the idea that in any development or conservation project the needs of the local people must come first. Long ago he began trying to redesign Kenya's national parks so that local herders could use, without over-using, the grasslands being reserved for wild animals; he also insisted that some of the tourist revenue should go directly to local people rather than into the national treasury. Where this has happened in Kenya, poaching has decreased dramatically.

The IPAL project, set up mainly to carry out research, officially ended in late 1983, and now Dr Lusigi runs the Kenyan Arid Lands Research Station at Marsabit, where he is responsible for putting into practice the lessons learned during the IPAL investigations. (With the demise of IPAL, research operations were handed over to the Kenyan government, whose activities are centred in the Kenyan Arid Lands Research Station. However, UNESCO and various development agencies are carrying out the recommendations of the IPAL Resources Management Plan, which appears to have the backing both of the Kenyan government and Northern aid agencies.)

Dr Lusigi blames the 'fast development approach' discussed above – 'big projects, big farms, settlements, big commercial centres, mechanisation' – for much of the desertification which has now affected the entire Sahel region. After the big, poorly planned projects were executed 'droughts came and wiped out whatever was left', he said.

'This area we are standing in now is a man-made desert,' said Dr Lusigi, looking out over the landscape around Korr. 'What has happened here is the result of introducing permanent water in a place where there should not have been permanent water at all. The people have been in here with their livestock and have overgrazed the area completely. They have cut down the trees. After the trees are gone, then the rain comes and strips the ground and the wind takes up the soil. That is the phenomenon we call desertification.'

IPAL not only 'integrated' its research, but integrated its prescriptions for saving the environment – many of which appear at first glance to have nothing whatsoever to do with the environment. One of the first things that the scientists realised was that they should avoid sentimentalising nomadism. Many Rendille were clearly anxious to settle. Trying to induce them all to return to the nomadic ways of life would have been a 'romantic notion', said Dr Lusigi. The goal is not so much to encourage all of the people to move again as to get their animals moving, giving over-used grasslands time to recover. But this does mean mobilising a reasonable number of herders in order to accompany the herds, and such mobility requires at least three things.

First, it must be physically possible. There must be enough wells scattered around the plains to enable the herds to move across those plains withour dying of thirst. Thus, many small wells are being dug. If the herds are mostly camels, rather than sheep and goats, they can cover more ground with less water. Camels are also less destructive to the soil and the vegetation. Scientists have therefore been encouraging the Rendille to return to their traditional camel-based herding.

Second, nomadism must be made less of a hardship than it is at present. UNESCO workers are trying to establish 'mobile shops' on the backs of the camels of *morani*, so that consumer goods such as sugar, tea, flour, tobacco and radio batteries can be taken out to the *fora* camps. These same camels bring into Korr the milk and firewood necessary to the family members who have been left behind, especially the children.

Third, the Rendille must believe that by herding they can realise their ambitions, whether these be the purchase of a pair of jeans or the longer-term hope of sending all their children to school. The Rendille will not revert to a nomadic way of life just because a UN agency tells them to; it is essential that they see it as a better option than depending on food handouts in town, or chasing the few jobs available in Nairobi. For this reason UNESCO has gone into the livestock herding business itself. It has also organised the Pastoralists' Association, one of which D'igir Turoga heads.

In the past the Rendille did not need to sell livestock because they had little use for money. Now, however, there are the shops in towns such as Korr, as well as the new mobile shops. School fees come to the equivalent of £50 ($75) per year, with the added expense of uniforms and materials – this in a country in which the average annual income equals only a few hundred pounds. When the herders first began selling their stock, they were cheated by 'middle men': traders who took advantage of the herders' commercial *naïveté* by buying cheaply in the drylands and then selling at higher prices in the animal markets of Nairobi. UNESCO has organised a monthly auction where it buys animals for sale in Nairobi. Each month a few members of the Pastoralists' Association take the livestock lorry to the capital, so that the herders can see how the Nairobi markets work. In doing this they are adding the skills of modern commercial dealing to their traditional herding and drylands survival skills.

Matters are helped by the fact that the main buying agent in Nairobi is a Rendille, also a member of the Association. But the Association is more than a club for the sale of livestock. In the old days, when each clan was gathered in a *manyatta*, there was a council of elders which met every morning to decide on the herding strategies of the day and of the days to come. As the *manyattas* were abandoned for the towns, this forum for exchanging information and passing on

The shop in Korr: it gives herders a need for money, which they did not have before

traditional knowledge vanished. The Association revives the practice, allowing for new learning in the process.

'It brings the pastoralists together,' said D'igir. 'And also it's good because there are new ideas which were not known before, especially where these animals are sold and where they are taken.'

After one of the monthly auctions, as bleating sheep were being thrown into the back of a lorry like sacks of grain, Dr Lusigi explained the advantages of having an auction on a given day in a given place. 'In the old days,' he said, 'one of the problems was the uncertainty of being able to sell an animal if a herder brought it to market. A herder might walk seven days to bring his animals to Marsabit, and might not be able to sell them for another seven days. If he didn't find a market, he had to walk back home with the animals.'

D'igir took up the tale: 'It was very difficult for people, especially single people. If they wanted to sell their animals they had to take them all the way to Marsabit and nobody was left behind to look after the other animals.'

The Kenyan pastoralist and the Kenyan academic agree on most issues; they see the same problems and, more important, the same solutions. However, what D'igir regards as issues that are particular to Korr and to Rendille land, Dr Lusigi sees as threats to the livelihoods of pastoralists throughout the Sahel and in other parts of dryland Africa. D'igir described how the local desertification is caused by humans and livestock staying near the Korr wells, but Dr Lusigi views the problem on a larger scale: 'The permanent borehole must be considered one of the biggest disasters that ever happened across the Sahel, because this environment is not meant to have permanent water that stays in a place and flows throughout the year. Abuse an ecological system and the result is death. If we cannot find a realistic solution, I see disaster and the continued starvation of millions of people in the Sahel for many years.'

Despite their very different educational backgrounds, the two hold similar views on the dangers of the type of schooling available to the herders' children in Korr, where the syllabus is exactly the same as that taught in Nairobi, and virtually the same as that taught in London. D'igir himself was one of the first of the children who were sent to school from his clan. At the time, the colonial administration was putting pressure on the elders in order to make more Rendille attend school. The elders told D'igir's father that he must send at least one of his three sons to school. (There was a daughter, but no one suggested that she should be educated.)

'My father was very sad, and the old men didn't want to offend him so they collected goats, and my father was given these goats – for sending one child to school. Yes, he was bribed!' D'igir laughed. 'I was really scared when I went to Marsabit. People wearing clothes – I'd never seen that before. And all those hills – I had never seen a mountain before. We used slates at school, and at first I didn't know what it was all about. I got things wrong and used to lick out my mistakes with my tongue.'

D'igir's five children range in age from four to thirteen, and he is struggling to educate them, both in school and on the grasslands. He keeps the two who are still too young for school with him and his wife at the *fora* camp, and he sees to it that the older children also do their share of the herding during the school holidays, as is traditional and as he himself did when he was a boy. 'Education is taking our children away. In the old nomadic way, the children used to come home [to the *manyattas*] during their holidays; but now living in settlements they don't learn much of their culture because they stay in towns.'

Dr Lusigi sees the same problem arising: 'As a result of inappropriate education, pastoralism is increasingly becoming an endangered way of life. The pupils who finish school are increasingly unsuited for living in the arid areas and never go back to pastoralism.'

The IPAL team pioneered a novel form of education for the adult herders, a technique which may be termed 'teaching by bad example'. UNESCO bought some camels and integrated them with the herd of camels belonging to an influential but poor elder named Lengima. A smiling D'igir recalled: 'They kept UNESCO camels and Lengima's camels together. Lengima's camels were treated [for pests], but not UNESCO's. Lengima's kept on growing, but UNESCO's camels had ticks, mites and diseases, and many calves died. In my opinion this was a way of

D'igir's son at school in Korr, adding modern to traditional skills

educating us. Because of the treatment, Lengima has a lot of camels now; last season he had about thirty calves. He doesn't stay around as a poor man in Korr. He takes care of himself and his family.'

The IPAL team have also used 'exclosures' for the study of natural processes in the grasslands. These are fenced *en*closures designed to keep livestock off a given area of rangeland and in order to see what happens to the vegetation when it is not subjected to the pressure of grazing. Perhaps more important than the data which these provided for the botanists and soil scientists, the exclosures proved to the herders that even the most desertified land was by no means ruined for ever. The land could recover if given a rest from teeth and hooves.

The 'Resource Management Plan' which IPAL initiated and which the Kenyan government and its Arid Lands Research Station have taken over is ambitious. It calls for the digging of 400 wells, one every ten square miles in the Rendille area alone. The cost of this would be about £800,000 ($1·2 million).

A start has been made, and in the process the contrasting methods of the traditional and the modern experts have been highlighted. D'igir Turoga travelled into the bush with a Californian hydrologist who took with him an expensive electronic seismograph. By using his eyes and studying the lie of the land and the vegetation, Turoga gave the verdict either, 'Here lies water' or 'No water here'. The hydrologist plugged his seismograph into the soil while his assistant banged a metal plate some yards away. As the reverberations through the ground were registered, he was able to take readings from the wave patterns on the device's oscilloscope. In every case the expert agreed with D'igir's eyeball assessment.

If the land is to be used efficiently, the Kenyan government will also have to take firmer measures against the banditry and inter-tribal raiding which still plague the region and put valuable grasslands out of use. Livestock raiding was traditionally a rite of passage for the young warriors. There were relatively few casualties in the days of spears, clubs and shields; wounds attracted the attention of young women and were something to boast about by the camp fire. With the advent of automatic weapons, a skirmish today can leave many dead. The situation is complicated by the fact that raiders may come from Ethiopia to the north and from Somalia to the east.

There are plans to bring in bamboo from around Mount Kenya to the south so that fencing can be provided without the herders having to cut down trees to make corrals. Hilltops to the east and west of the Rendille flatlands must be reforested to protect watersheds and to keep rains from becoming destructive floods. Co-operative livestock management systems dividing up rangelands among herders must be set up and enforced by local range-management officers, who must be hired and trained. Dr Lusigi wants to see savings and loan banks in the district, and training provided so that the Rendille are able to learn how to manage money as well as camels.

These measures may seem to involve a great deal of trouble and expense just to keep a relatively small number of people gainfully employed in a hostile landscape. But the Kenyan government has strong motives for wishing to carry out the work. Kenya consumes and exports more meat than most African nations, and well over half of that meat comes from the drylands. Ultimately, the success of IPAL and similar efforts made by dryland pastoralists may be crucial to human welfare and financial security throughout the African continent.

Africa is home to some 15 to 24 million pastoralists, most of whom live in the semi-arid and arid regions which make up three-quarters of the continent's land area and are where most of its meat is produced. Nations such as Sudan, Somalia, Chad, Ethiopia, Kenya, Mali and Mauritania each have 1 million or more herders. Nomadic herding is far more efficient than any commercial ranching system yet devised for the African drylands, and studies in the Sahel have found that traditional herding methods produce more protein per acre than do the ranching systems of semi-arid Australia.

In the late 1970s, UNEP came up with a huge scheme to take advantage of the meat and milk being produced throughout the Sahel. The aim of the programme, known as 'SOLAR' (Stratification of Livestock in Arid Regions), was to turn the Sahel into the 'meat-basket' of the world. The idea was that nomads would rear calves in the traditional way; the calves would be moved south and fattened in the wetter farmlands or on feedlots, and finally sold in the prosperous states of coastal West Africa, much of the beef being earmarked for export. SOLAR was a disaster, a failure attested by the abandoned abattoirs and feedlots that dot the Sahel. No one had thought of asking the nomads whether they had any interest in selling calves and no attempt had been made to motivate them to do so. The planning behind SOLAR was based on what the planners wanted rather than on what the herders wanted.

IPAL had taken a completely different approach. As Dr Lusigi said: 'People are the key to solving the problem of desertification. And the only way we can be able to suggest any new approaches is to learn to understand the culture of the people, their thinking, their expectations. What do they think they want for themselves? What is their thinking while over-using their environment?'

It would be wrong to suggest that this people-based philosophy guarantees success. On the contrary, there are many problems involved. The Rendille complain that UNESCO is not paying them enough for their livestock at the auctions. The complex camel transport arrangements to convey consumer goods to the *manyattas* and milk back to the towns are not working well. The price of livestock products is kept artificially low by Kenyan government policies, so the herders are not as keen to sell as they might be in a freer market. Dr Lusigi admitted the problems, but he insisted that the people-based approach offers the only hope of success.

Many of the Rendille themselves are beginning to feel the despair of living on grain handouts in Korr and other towns. They now realise that they must make a basic choice between either devoting themselves to better, more profitable herding or abandoning everything for the unlikely chance of a job in Nairobi.

'A lot of people are wondering about me in particular,' D'igir Turoga said. 'Sometimes I go to Nairobi, either as an Association member or on my own. Sometimes I dress differently. So people are wondering how I am so adaptable. They are asking me why I am trying to stay the way I am.' D'igir thinks he sees a way of thriving, of becoming 'rich', as other Rendille herders have done. He remembers the lesson of how Lengima's camels were made healthy by veterinary treatment, and how Lengima is prospering now.

D'igir maintained that animals 'are the past, present and future of this place. Pasture is there in Rendille country. It is big country, which even now is not used up.' He is teaching his children how to keep themselves and their animals alive on the plains. He is also giving them a modern education, and he wants at least one to become 'a human doctor or a veterinarian'.

Other dryland herders in other places have realised such ambitions. In Texas, some poor, young, tough, well-armed nomadic herders – known as 'cowboys' rather than *morani* – grew rich by moving livestock efficiently across the dry plains. They educated their children and many of the children stayed in the cattle business from generation to generation. Others left to seek success in cities. That, after all, is the nature of the development business: wider choices and expanded possibilities. A Rendille should not be forced by circumstances beyond his control to live in Korr relying on handouts. Neither should he be forced to move to the alien shanties of Nairobi, or to remain a pastoralist.

Dr Lusigi agreed: 'I think the most urgent thing that needs to be done in the pastoral areas is to strengthen the pastoral economy, to make the pastoralist a strong person economically, to be able to survive modern influences that are pushing him to the brink of the world. He will need to climb back up the ladder, to where he can compete with the rest of the economy.' Only then, 'can the pastoralist choose what he wants to do'.

As for D'igir Turoga, he has seen it all: the bright lights of Nairobi, the security of serving in the police force, and the town life in Korr. And he has made his choice: 'The pastoral life is very hard, but we like it. We depend on these animals. That is how we get food, how we get our living, how we get everything.'

V

SOLOMON ISLANDS: FISHING THE COMMONS

'I believe that the bold stand taken by my government has really strengthened the very nature of our fishing industry. We have our rights, our 200-mile limit, and if anyone wants to come and fish, then they must do it in the right way.'
MILTON SIBISOPERE, head of the Solomon Islands' National Fisheries Development

According to Dudley's mother, her son fitted neatly into the Northern world's stereotype of a typical South Pacific islander: lazy, spoiled, living easily off the bounty of nature.

'Dudley was a spoilt child,' admits Mrs Anna Tapalia. 'He was the youngest of seven children and always insisted on having the best of everything, whether food or clothing or other things. He used to spend all of his time in the bush – bird hunting, diving in the river or just strolling along the beach in order to be out of sight of his parents and not be told what to do. Early he would leave his house: late he would come home.'

Dudley's village of Aola, on the north coast of the island of Guadalcanal, recently connected by road to the Solomon Islands' capital of Honiara to the west, is a paradise for children who like to hunt and fish. The village consists of a collection of houses situated on the edge of the beach. Most of the dwellings are made in the traditional way: light poles support rafters above which are roofs of sago palm leaves; the walls are bamboo or thatch. A river enters the ocean near by, adding freshwater fish to the village diet. Behind the village is a forest – 'the bush' – which is abundant with birds and small game. Most of the villagers own a log canoe, and in the mists of a picture-postcard dawn youths can be seen canoeing around the shallows catching the evening meal, a sharp-pointed paddle in one hand and a multi-pointed spear in the other.

Dudley Tapalia

Dudley Tapalia's childhood seems to have agreed with him. He has grown into a big, muscular youth of nineteen. His dark face contrasts with healthy white teeth produced by a diet rich in fruit, garden vegetables and fish and low in sugar. His daily costume consists of shorts and a T-shirt, a garb which has become the 'national dress' of the Solomon Islands. The most sought after T-shirts are those bearing the names of Western cigarette brands, cars and pop groups.

When asked why he likes his current job on a tuna boat Dudley's apparent indolence persists: 'Village life is terrible because there is lots of work that I cannot handle and I don't have any rest days in the week.' He says that he left his job with the South Korean Hyundai Timber Company in Aola after only eighteen months because the office was a long way from the sawmill and he had to do too much running to and fro. He likes life at sea 'because that life is free and enjoyable. We get the best food: rice and corned beef and even tea with milk. There are about twenty-eight boys on the boat and we have a lot of fun.' With more training Dudley hopes to progress from general crewman to 'spymaster', the officer in charge of spotting schools of tuna. 'The spymaster's job is very easy; he just sits on the top of the bridge with his binoculars to find fish. He only works during the day and sleeps at night when the rest of the crew is fishing for bait fish.'

Despite his public image of the lazy islander, a persona which he deliberately seems to adopt when dealing with foreigners, Dudley is in fact a very complex, active and ambitious youth. His complexities and ambitions, indeed his very youthfulness, in many ways reflect the complexities and youthfulness of the Solomon Islands nation.

The Solomon Islands, which gained independence from Britain as recently as 1978, comprise a banana-shaped line of six large islands and innumerable smaller ones curving south-eastwards from Papua New Guinea and running parallel to the north-east coastline of Australia. This banana is split by 'The Slot', a deep trough abundant with tuna and other fish which runs between the main islands. Indeed, most of the waters now controlled by the Solomon Islands in their 200-mile Exclusive Economic Zone (EEZ) are rich in tuna and marine life. The territory enclosed by this Zone amounts to 38,600 square miles of sea, an area almost as big as East Germany and four times greater than the islands' land area.

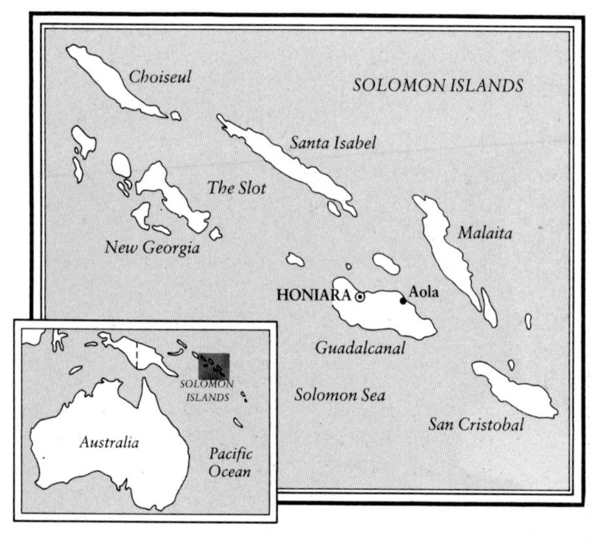

On the land itself lives one of the most ethnically and socially diverse mixes of races and customs as is anywhere contained in such a small space. The majority of the islands' 260,000 inhabitants are Melanesian, whose people are also found on New Guinea and other nearby islands; but there are also a smattering of lighter-skinned Micronesians from islands to the north, Polynesians from archipelagos to the east and Caucasians from Europe and North America. The term 'Melanesia' comes from the Greek meaning 'dark islands', but is thought to refer to the dark greenery of the slopes or the sombre storm clouds often hovering over the islands rather than to the colour of the people. During the first few hundred years after a Spanish vessel became the first European craft to land on the Solomon Islands in 1568, they were known – by those who had heard of them – as the homes of cannibals and headhunters who raided one another's islands in elaborate war canoes and were liable to kill any Europeans who landed. (Modern anthropologists argue that the islanders ate the flesh of vanquished enemies not as a source of protein or as a delicacy, but as part of a ritual to gain spiritual power.)

Charles Woodford, who became the first resident commissioner when Britain proclaimed its protectorate in the Solomon Islands in 1893, added to the islands' reputation for savagery in his book *A Naturalist Among the Headhunters*. He gave descriptions of the natives wearing either nothing at all or a small cloth around the waist, enlivening his narrative with such titbits as: 'I was told sometimes that men had gone into the bush to hunt wild pig, but it was really to kill men, but the men of Guadalcanal were never cannibals as in the other islands.'

As recently as 1920 the British Foreign Office had found little improvement in the islanders. An official report said: 'The natives are restless and warlike, addicted to headhunting and cannibalism; and on many larger islands, especially Malaita and Choiseul,

Aola: traditional houses situated between the forests and the sea, both of which sustain the villagers

there is incessant warfare. Towards whites they have the reputation of being cruel, treacherous and revengeful.' Even now the Solomon Islands as a nation and the islanders as a people have a reputation for fierce independence and an unwillingness to be controlled by foreigners.

Today 200,000 islanders continue to live in villages of fewer than 200 people. They have little contact with the government and ministries in Honiara; less than 10 per cent of the population are listed as being formally employed. The character of the villages and the villagers varies tremendously from one steep, forested island to another. In coastal villages, such as Dudley's home of Aola, the inhabitants live off garden produce and the ocean's protein; many live over the sea itself in palm-thatched huts on stilts. These so-called 'salt water people' have completely different life-styles from those 'bush people' living only a few miles inland on the same island. Having no ocean on their doorstep, the 'bush people' raise pigs and grow yams, taro, papaya, bananas, nuts and coconuts. The seaside and inland villagers often cannot even talk to one another, as over sixty languages and dialects are spoken in the islands as a whole. The common denominator language of 'pidgin' grew spontaneously and spread to allow communication between the islanders. The official language of English is now beginning to rival pidgin as more of the islanders go to school.

Dudley's sixty-year-old father, a self-appointed Methodist pastor in Aola, lived most of his life during a time when, as he remembered, 'money had no value; we planted food for our living, unlike now when you have to pay for all these things'. In fact,

Dudley (top right), *with his parents and other relatives, listening to stories in the evening*

traditional society frowned on the accumulation of wealth, whether in the form of land or 'money' – red feathers, porpoise teeth, bat teeth or strings of shells. In the eyes of the village, men became 'big men' if they helped their clan or their village by sharing their wealth and holding feasts. If a man was too openly greedy, his fellows might dub him 'Mr Me' or remind him of the transient nature of material wealth by reciting the taunt: 'Young man, rich man, dead man.'

Dudley's mother Anna, a thin, white-haired woman with a fierce scowl and decided opinions on how young men should behave, is also a powerful force in the family. On this part of Guadalcanal, as on some other islands, inheritance is through the mother's, rather than the father's, line. Dudley thus owes as much allegiance and respect to his mother's brothers as to the man he refers to as his 'true father'.

As the need for money grew in the islands, whether for taxes, school fees, radios or outboard motors, the islanders began to combine their wholly subsistence life-styles with the selling of fish in the bigger market towns and the sale of copra (dried kernels of coconut) to buyers for export to Japan, the USA and Europe. The new island government also found itself in need of cash for textbooks, medicines, roads, boats, planes and other means of communicating with its far-flung citizens on their 870-mile-long archipelago. In terms of natural resources, the islands had three main commodities to offer the world: timber, fish and copra.

It is easy enough for a developing nation to make money from its timber. All that must be done is to invite in a company to fell the trees, for which privilege a fee is charged. This is precisely what Dudley's eldest brother did when he negotiated with the

Inside the sawmill at Aola

Hyundai Timber Company of South Korea to allow it to cut down trees near the village. Two of Dudley's three elder brothers still work in the timber business and the eldest, now living in Honiara, has become a very wealthy man.

Such logging ventures have become a burning issue in the islands, with the Ministry of Finance trying to maximise revenue and the Ministry of Natural Resources trying to limit the felling to a sustainable, non-destructive level. If a small nation sells large quantities of timber for too long, supplies will simply run out. On the steep islands, where tropical rainforests grow in mostly poor soils, logging can not only quickly destroy the forestry but can also rapidly erode the bare hills. During rainfall the soil of village garden plots is washed into the sea, leaving the villagers with neither trees to sell nor soil to till, also killing reefs where fish breed. The villagers of Aola who lay claim to the forested land through village customs of inheritance and ownership, rather than by modern deeds, are divided on the issue. Many are in favour of the sawmill and the Koreans because with the industry comes employment in the forests and the mill, and they are able to make furniture for sale from the leftover wood. Other villagers feel that they are not receiving their fair share of the proceeds from the logging. On several occasions the offices and living quarters of the logging camp have been vandalised and the company furniture destroyed.

It is also easy for a Third World nation to make money from its fish. Again, all that must be done is to invite in companies from the big fishing nations and charge a fee for fishing rights. Many Southern coastal nations either lost patience with drawn-out UN Law of the Sea negotiations and proclaimed their own 200-mile zones, or had such zones in effect thrust upon them at the end of the UN negotiations in 1983. However, because many of these nations lack their own deep-water fishing fleets, they have followed the path of least resistance by issuing fishing licences to foreign fleets. If a nation is so poor that it cannot afford the aircraft and boats necessary to monitor such fishing, foreign 'distant water' fleets tend to ignore the territorial boundaries, taking the fish and not paying the fees. The picture is further muddied by the fact that by 1986 it was clear that it would be many years before the Law of the Sea treaty took effect, mainly because not enough countries had ratified the agreement but also because nations such as the USA, Britain and West Germany had rejected the treaty. Agreements, and disagreements, over distant water fishing have raised issues that the world has yet to resolve.

For instance, the waters of the Atlantic Ocean off the drier nations of West Africa are so rich in fish that the recorded catch before processing is worth £930 million ($1395 million) each year. However, most of this catch is frozen and exported to Europe, Asia and other parts of Africa. Little of the profit finds its way into the treasuries of the coastal African nations, less still to the people of those countries. But these and all other tropical coastal nations do possess fishing fleets of a sort; fleets of men in canoes and small boats who, like Dudley's fellow villagers, go out each day with hooks and nets to catch fish in the shallow waters and sell them locally. Throughout the Third World there are about 12 million of these so-called 'artisanal' or peasant fishermen fishing full-time, approximately twice as many fishing part-time, and another 12 million people – many of them women – involved in supporting industries such as cleaning, curing and marketing. Several hundred million people in the South depend on fish for almost all their animal protein; fish supply over three-quarters of this protein in nations such as Bangladesh, Burma, India, Indonesia, Sri Lanka, the two Yemens and Trinidad and Tobago.

From 1950, when the world's fishermen landed 21 million tons of fish, until the early 1970s, when they netted over 70 million tons, the potential increases in the catches seemed boundless. Since those years, however, the ocean harvest has stagnated, having crept up to only 76 million tons by 1983. Northern fisheries of such species as herring and haddock were grossly overfished, and nations such as the USA, the USSR, Japan, China and South Korea started to look far outside their own depleted waters for their catches. Distant water fleets have begun to compete directly with the peasant fisheries. Factory fleets catching shrimp in the shallow waters off India scoop up not only the shrimp sought by the local fishermen but many other species as well, the latter of which may be thrown out as 'trash fish'.

It is not only foreign trawlers that provide competition for the local peasants: in 1980, trawlers both from coastal states and from further afield were catching over 1 million tons of fish per annum in the Gulf of Thailand, an area in which the maximum sustainable yield of fish has been estimated at half that quantity. There have been several armed clashes between trawlers and peasant boats off both Malaysia and Thailand, including shoot-outs, deliberate rammings and boats being set ablaze. The issue has become so acute that the UN has been holding special meetings on the needs of 'small fishermen', and peasant fisheries were made the theme of the UN Food and Agriculture Organisation's 1986 'World Food Day'.

Small and isolated though they are, the Solomon Islands have not ignored such trends; neither have they failed to learn from them. As early as 1971, they began what has been a long, delicate tightrope act – balancing on the one side the desire to make money quickly from their stocks of tuna and on the other the need to husband this resource in such a way that it would provide maximum benefit for the greatest number of islanders over the long term. One of the first facts that the islanders took to heart was that, according to studies, if a maritime nation simply licensed its seas to foreign fishers the most it could hope to earn was 5 to 10 per cent of the total 'first sale' (before processing) value of the catch.

In addition, six out of every ten Solomon Islanders are, like Dudley, under twenty-one years old. Many of these will soon be starting families, and the islands' population is already doubling every twenty years. At the present rate of growth the government will be seeking to provide 50,000 new jobs for people such as Dudley by the mid-1990s.

In 1972, after a year of negotiations, the Solomon Islands signed a ten-year joint venture agreement with what is perhaps the world's largest fishing company, Taiyo Gyogyo (Fisheries Corporation) of Tokyo, an organisation with diverse, worldwide investments and whose annual sales of over £2·3 billion ($3·5 billion) total more than twenty times the Gross National Product of the Solomon Islands. There have been many such partnerships between Third World nations and Northern companies, few with very much success. Tony Hughes, a former British colonial civil servant in the Solomon Islands who married an islander and stayed on to become governor of the Central Bank, explained the nature of the problem: 'Joint venture is not an easy ball game at all. People often talk about joint venture as if somehow the interests of the foreigner and the host country are the same. They are never the same. But, if there's a sufficient overlap, the joint venture can exist in that overlap. And you have to make sure the overlap continues to exist.'

What Taiyo Gyogyo wanted was guaranteed long-term access to one of the richest populations of skipjack tuna close to Japan. Events over the years have given the company several reasons for being pleased with the joint venture. During the 1970s, demand for canned tuna increased sharply in the industrialised world as fresh fish became scarcer and more expensive. This coincided with many newly-independent island nations making declarations of EEZs, the advent of which suddenly increased the island nations' oceanic 'territory' tenfold and left very little fish-filled free water in the Pacific. Japan, forced to go further afield for tuna, needed live bait fish for its boats which fish with poles and lines. It is easier to catch such fish near the tuna grounds than to bring them from Japan. As these fish are caught on the reefs near the shore, the Japanese company needed firm agreements with the reef-owning nations, EEZ or no EEZ.

The Solomon Islands saw in the deal with Taiyo Gyogyo a way of realising several of their own goals: a share in the profits from tuna sales; an entrée into world markets offered by one of the few companies that could compete with the other big international companies; the development of a Solomon Islands fishing fleet and employment for islanders on boats and in canneries and smoke houses.

The original joint-venture company, called Solomon-Taiyo Limited, was an extremely complicated partnership combining Japanese capital, equipment and management with the Solomon Islands' manpower, fishing rights and fish. During the first ten years of the venture the Japanese parent company had a 75 per cent interest while the Solomon Islands had 25 per cent. In a new partnership which took effect in 1983, each side owned 50 per cent with the proviso that the Solomon Islands' share would increase to a controlling 51 per cent after five years. Towards the end of the 1970s, when tuna prices were at their highest, the Solomon Islands were earning £4 million ($6 million) per year from skipjack tuna, and the fishing industry accounted for almost one-third of the value of all exports.

For Dudley Tapalia and other young Solomon Islanders the world was changing with the same speed as had the world of the Rendille tribesmen such as D'igir Turoga (see Chapter Four). It was not that the islanders' way of life had been led into an environmental dead-end as was the case with the Rendille herdsmen. Gardens and forests remained productive; the seas still teemed with fish, and the islanders knew how to harvest the oceans for their own consumption. One New Zealander, who was sent to the Solomon Islands by the British Overseas Development Administration in the late 1970s to teach the western islanders how to fish, said he found his technical training useless: 'Their knowledge of fish was spectacular; they had been catching them for thousands of years; all I could suggest were things like plastic nets, which forced them into the money economy. I felt redundant.'

He found men on Ranongga Island flying sago palm-leaf kites out over the edges of the reefs. Each string which formed a kite's tail had on the end of it a wad of very sticky spider's web gathered in the forests. As the web skimmed over the water, garfish would bite it, become entangled by their teeth and be hauled in. Elsewhere islanders in canoes had already anticipated the tuna-catching method using barbless hooks which the Japanese later 'introduced' to the islands. The method was to file a seashell into a fish-like shape and attach a barbless hook carved from turtle shell to this lure. When the tuna bit, it was flipped into the canoe by the fisherman holding the line.

Dudley's own people used various techniques, one of which was the *visi* method. This involved making a mesh of coconut leaves and vines which was held underwater by men surrounding a school of fish which had ventured into the shallows. The fishermen shouted and rattled the fronds in order to frighten the fish and the 'beaters' closed in on the school, penning it in against the shore. Canoes were placed parallel to and a few feet behind the line of beaters. When the fish felt trapped they would leap over the mesh and, more often than not, into the bottom of a waiting canoe. These techniques caught sufficient fish for the villagers but not large enough quantities for export. Those that were caught were rarely sold but would be divided on the complicated basis of family and clan obligation and entitlement which dominates village life.

The *visi* method, like so many old techniques and customs, has died out. The culture of the Solomon Islands remained intact during the 'blackbirding' or enforced labour period from 1860–80, when Melanesians were captured in raids and taken off to work in the sugar fields of Australia and Fiji. It also survived the killer diseases such as influenza, malaria, tuberculosis and others which were inadvertently introduced by European visitors in the early 1900s. It held out during the year-long Japanese occupation in the Second World War and throughout the bloody fighting on Guadalcanal which is thought by many to have turned the tide in the Pacific campaign. (It was a Solomon Islander, Ben Kuvu, who with seven companions in canoes rescued the US President-to-be John Kennedy after his PT Boat 109 was rammed by a Japanese destroyer.) Yet the Solomon Islands' culture has been slowly succumbing to what may be one of the most powerful forces on earth: the desire for modern ways and modern manufactured goods.

'We could teach the young people our culture and other things, but who are we going to teach?' asked Obed Suba, an elder of Dudley's village. 'Most young people won't listen; they think they know better and tend to drift to town and other places in order to enjoy this "modern civilisation". It is Western influences that attract their attention, and they tend to forget all their culture and customs and pretend to be clever and forget to listen to their elders.'

All over the world parents complain that their children think they are too clever to listen to their elders and betters. The difference is that when a boy leaves his father's farm in Kansas to seek his fortune in Los Angeles, he remains firmly rooted in the same basic US culture, but when a boy leaves Aola to map out his future in Honiara, only thirty miles down the coast, he makes a trip covering thousands of years. He will be wrenched from a society which judges a man by how he fulfils complex obligations to family, clan, friends, village and ancestors into a society which judges a man by salary, job, clothes and possessions.

Dudley has made the transition gradually and in fact has been helped by changes in the village itself. When Dudley left secondary school at the age of sixteen after his third year, he started his own cocoa plantation on family land, a plantation which is now earning him money from the sale of cocoa. When Dudley worked at the South Korean sawmill with his brothers and other villagers during the two years before he joined the tuna fleet, he rose quickly in rank and responsibilities. He was soon calculating and co-ordinating export orders for the manager, another example of his closely-guarded enterprise and ambitions. However, Dudley wanted to see more of life on his own and not always be under the thumb of his Korean bosses, his parents and his elder brothers. One day, he heard a radio advertisement for vacancies in the tuna fishing fleet. Without consulting his parents – 'they were cross with me, but I am a free man and can go where I want' – he went for an interview in Honiara and convinced the personnel manager that he was not afraid of the ocean when it was rough and was not prone to seasickness. Dudley was told to go back to his village and listen regularly to the radio which, in the Solomon Islands as in much of the Pacific, serves as newspaper, mail, telephone and telex. In a week or so Dudley heard a 'service message' asking him to report to his ship. He had become a modern tuna fisherman.

Dudley did not go to work for Solomon-Taiyo Limited, but he joined the 'catcher boat' *Solomon Pathfinder* belonging to the National Fisheries Development, a more recent example of Japanese–Solomon Islands co-operation. Although the Solomon-Taiyo agreement was designed to be a joint venture, the government of the Solomon Islands felt in the mid-1970s that its people were not benefiting from it as they should. After all, the Japanese company was in it for the money; its sole aim was to catch the maximum tuna for the minimum investment. The company therefore filled its boats with fishermen from a small group of islands south of the Japanese island of Okinawa. These men, although culturally different from the people of the big island to the north of them, are generally referred to as 'Okinawans'; they have been fishing the Pacific in family-owned or leased boats for decades and are extremely good at it. Solomon-Taiyo tended not to train islanders in advanced skills, arguing that the men often stayed on the boats only for a year or so until they had made some money and then they would return to their villages and their former ways, using the money saved to pay bride prices where that custom pertains and perhaps to buy some land. Furthermore, whereas the Okinawans were used to labouring round the clock, fishing for tuna by day and for bait on the reefs by night, the islanders preferred work-patterns which allowed time off for socialising. Not only was there no opportunity to socialise on the boats, but the islanders could communicate with the Okinawans only by sign language, a medium that made the learning and appreciation of the different cultures impossible on both sides. In the early days of Solomon-Taiyo, there were frequent battles using fists and clubs between islanders and Okinawans on the boats, and a Japanese engineer was stabbed to death at the company's cannery.

When it set out to develop its own fishing fleet and train its own people, the Solomon Islands government established the National Fisheries Development (NFD) in 1978, with 25 per cent ownership and the loan of two boats by Taiyo Gyogyo, and loans from the Asian Development Bank to build ten ferro-cement tuna boats. The ferro-cement boats are not as durable as steel boats, but neither are they as expensive. More important, they could be built by the Solomon Islanders. The NFD has organised itself more on the basis of the cultural traditions of the Solomon Islands. For example, by allowing the fishermen more time off so that they can return to their villages, staff turnover has been reduced. When the calls go out on the national radio for fishermen to return from holiday to their boats or otherwise lose their jobs, many more NFD fishermen return than do those working directly for Solomon-Taiyo.

Dudley joined an NFD boat with twenty-eight crewmen, only one of whom, the fishing master, was an Okinawan. Not surprisingly, Dudley did not like the master: 'He gets cross so much with the men and it would be better if we could get rid of him, get a Solomon man to take his place.' Dudley was indirectly referring to the issue of 'localisation' or replacing Okinawans and Japanese with islanders. This issue brings up the very essence of the delicate balance which the Solomon Islands are trying to achieve for their fisheries and their nation. The reason why Okinawans cannot quickly be replaced with islanders is that Okinawans know how to catch large quantities of fish. Before they are ready to act as captains, bosuns, engineers and spymasters, the islanders must gain the necessary knowledge and experience, and it would be much better for the economy of the islands if the fleet was making money while this was happening.

The type of craft which Dudley joined – a 'pole and line' boat – is also a part of that balancing act. It is the type of tuna boat on which the greatest number of people are employed. Furthermore, 'pole and line' tuna fishing provides the excitement which often appeals to a nineteen-year-old fisherman. When the spymaster spots a school, he rings a bell and the fishermen pour out on deck, grabbing their fibreglass poles and loosening the lines, each of which has a lure and a barbless hook attached to its end. The men line themselves shoulder to shoulder along the railings at each side of the boat, and sometimes the stern, with the best fishermen given the best positions. Finger-sized bait fish are brought up live from the boat's hold and thrown continuously over the side. The bait fish try to swim back under the boat as it is the only shelter in sight. In this way they attract the tuna to the boat, as do the nozzles on the boat's sides which spray water into the sea and thus mimic a bigger school of small fish in flight. When the tuna approach the boat they are feeding in so frenzied a fashion that they bite the shining lures and are flung over the fishermen's shoulders. A skilled fisherman can flip a tuna off his hook even before it hits the deck; he can then throw his lure back into the churning waters ready for the next victim. (On one occasion during only two hours of fishing from one school alone, Dudley and his comrades caught fourteen tons of tuna.) The deck itself soon becomes a mess of blood, slime and writhing, forearm-length tuna. The fish continue to cascade over the fishermen's heads for as long as the school remains near by and feeding. However, a man may be hit in the face by one of the fish or receive a hook in his eye if his neighbour on the rail is not properly skilled.

When the school stops feeding and disperses, the catch is gathered and stored on ice in the boat's holds. Dudley and his colleagues hose down the decks, repair the tackle, store away the rods and then wait for the next burst of activity. Periodically the boats visit the cannery, where they offload as quickly as possible, take on more food and fuel, and then cast off again, perhaps spending only thirty minutes at the dock to minimise the queuing time for the other boats. A fishing master wants to spend his time fishing, not tied up ashore.

At night the *Solomon Pathfinder* and the rest of the fleet anchor near reefs, lowering fine-mesh nets to the ocean floor and suspending a light in the water to attract small fish. As dawn approaches, the light is dimmed so that the fish are concentrated over the nets. The nets are raised and the bait fish for the next day's tuna fishing are thrown into the tanks on board.

The tuna processing plant: boats race here to unload fish and load on supplies so they can fish again

The villages which 'own' or have traditional rights to the bait-fish reefs are paid for the bait fish taken in this way. This provides further income for the islanders but leads to endless arguments and complications because it is never quite clear how to divide the money fairly among the villagers.

By employing local fishermen rather than Okinawans or Japanese, the Japanese company can cut its wage bills. Wage bills may be further reduced by abandoning the standard pole and line and adopting methods that require fewer fishermen. The Japanese have developed a pole-and-line boat with mechanised poles which reduce the number of crew required. There is also the purse seiner, the technique favoured by most of the big distant water fleets. A mile-long, 450-foot-deep net encircles a school of tuna, and a line is drawn in to close off escape through the bottom. In this way an entire school may be netted using winches rather than manpower. However, such capital-intensive approaches do not fit the Solomon Islands' purposes; neither employs many men and both use equipment the islands cannot afford. Therefore the government has disallowed the Japanese and the NFD from using either method. (Tuna are also caught by 'long-line' boats which deploy a buoyed line supporting smaller lines and baited hooks which stretches for perhaps sixty miles across the sea, allowing the bigger, man-sized fish which swim deeper to be reeled in. This technique has been used by the Japanese and the NFD around the Solomon Islands.)

The controversy over fishing techniques extends even to auxiliary equipment such as sonar and other aids which facilitate the detection of fish. The Japanese and American distant water tuna fleets carry devices which give them satellite-generated readings of surface seawater temperatures over a large area of the Pacific, an important clue as to where the schools are likely to be. When the Solomon Islanders want to check the water's temperature, they test it by using their fingers. They cannot afford to buy or maintain sophisticated electronic equipment. Neither do they yet have the skills to repair and effectively use such systems. On the other hand, if they ignore all technological developments their fleets will have difficulty in competing financially with the fleets of modern, more sophisticated vessels.

This Catch-22 situation is true of industries throughout the Third World. The biggest short-term profits come from using the latest, most capital-intensive equipment, whether in fishing, logging or harvesting pineapples. Thus governments of developing countries are tempted to do no more than license out such activities to Northern companies and take a share of the profits. Almost everyone praises the Solomon Islands' actions in boldly deferring income in favour of long-term development which will give the islands their own fishing industry. Yet fierce debates rage over whether or not they have struck the right balance. Many feel that the ferro-cement boats were a mistake, and indeed NFD officials admit that these smaller boats were 'outmoded' almost before they were built. Others feel that the NFD must start using purse seiners in order to remain competitive, and indeed it has decided to buy two such boats but limits may be set on how close to the islands they can fish.

The Solomon Islands' delicate calculations have been further upset by a collapse in world tuna prices which took place in 1981 and still persists, having led to the closure of all major tuna processing plants on the US mainland. Part of the trouble is simply that too many tuna are being caught, with American Tunaboat Association purse seiners increasing their catch from 90,000 tons in 1980 to 370,000 tons in 1984. The USA has not agreed to the UN Law of the Sea treaty (although it has proclaimed its own 200-mile EEZ), and a US law, the Magnuson Act, declares that tuna is a 'migratory fish' not belonging to any one country. Under this law the USA is permitted to place an embargo against imports from any nation which it considers has interfered with US tuna boats.

With prices falling, US boats have been under intense pressure to catch ever more fish. Some observers in the Pacific feel that the US boats have been taking what they could wherever they could, paying 'no regard to ecological or political consequences of fishing', in the opinion of an editorial in the *Pacific Islands Monthly* magazine which elsewhere has used such words as 'arrogance', 'insensitive' and 'pig-headedly selfish' to describe the attitude of the American Tunaboat Association. In 1984, a Solomon Islands' gunboat confiscated the California tuna boat *Jeanette Diana* found illegally fishing in the islands' EEZ. The USA responded by imposing economic sanctions against the Solomon Islands.

Philip Muller, a Samoan, is head of the Forum Fisheries Agency, an organisation set up by the South Pacific nations to license and monitor tuna fishing in the region. In his view: 'The sight of the big USA coming down with economic sanctions on a fairly small and very fragile community like the Solomons probably set the whole political scene against the USA. There was a backlash in the entire South Pacific. The Americans themselves say that sanctions never work, and all they got out of it was a lot of loss of goodwill.' The extent of that loss was brought home to Washington in 1985 when the tiny nation of Kiribati (population: 63,850), situated to the north-east of the Solomon Islands, shocked the US government by granting USSR boats licences to fish in its waters. Subsequently in June 1986, Vanuatu, south-east of the Solomon Islands, became the first small Pacific island nation to establish diplomatic relations with the USSR. The two nations immediately began negotiations toward a fishing agreement. Suddenly, the US attitude toward a renewable natural resource had become of importance in the East–West power balance in the Pacific. At that time the USA had no agreements to fish in any of the independent Pacific nations' EEZs, and negotiations for such agreements were still going on in 1986.

This jockeying for position by both domestic and foreign fishing fleets raises the issue which is always raised when 'renewable' environmental resources are being exploited, be they fish, trees, groundwater or topsoil: are the resources being used up too fast? A programme to tag and monitor skipjack tuna was begun in the late 1970s, one of the biggest operations of its kind ever undertaken on any wild species. The results revealed little more than that there are a lot of skipjack; they migrate vast distances; and there are no clearly differentiated 'local' geographical stocks.

Given that tuna do not respect national EEZs, if the Solomon Islands are to keep their own industry healthy they must join forces with the twenty-two other developing island nations of the Pacific in controlling the distant water fleets. The Forum Fisheries Agency, with its headquarters in the Solomon Islands and as representative of thirteen of those nations, gives the islands a force by which to exert such control, and its efforts have been widely praised around the fishing world. Philip Muller himself admires the way in which the Solomon Islands are fighting to conserve stocks while developing their own industry: 'It is not really that they are more vigorous in conservation than some of the other nations, but they feel they have more to lose. They've got an industry now that probably leads the country in terms of revenue and export earnings, and they are very conscious of this.'

Milton Sibisopere, who heads the Solomon Islands' National Fisheries Development, agrees: 'I believe that the bold stand taken by my government has really strengthened the very nature of our fishing industry. We have our rights, our 200-mile limit, and if anyone wants to come and fish, then they must do it in the right way.'

Tony Hughes of the Solomon Islands' Central Bank, also chairman of Solomon-Taiyo Ltd, is both proud and realistic: 'The fishing industry is in dire financial straits because of low tuna prices. It is only still there because it is owned by the government and Taiyo, both dedicated to the fishing industry. If it had been owned by a biscuit manufacturer or something, it would have been abandoned by now.' He adds that the Solomons-Japanese joint venture was a successful example of how rich and poor nations could co-operate in developing an industry based on environmental resources: 'It is good news for everybody.'

Meanwhile Dudley, when he is not dreaming of becoming a spymaster who is allowed to sleep while the rest of the crew are catching bait fish, hesitantly admits to wanting to be chosen for sponsorship at the technical college in Honiara in order to study marine engineering. 'When you finish that you've got an engineering qualification and you can go and work on any ship,' he said. For the moment, however, he is happy with the free life, fun and good food on the tuna boat.

He is happier still that there are several *wontoks* on the boat with him. *Wontok* is pidgin, literally meaning 'one talk', referring to people who speak the same language or dialect. It has come to be a metaphor for kinship and the sharing of mutual goals and values in much the same way as the word 'brother' is used in English. *Wontok* is often used to refer generally to the kinship and obligations systems which dominate so much activity in the Pacific island nations. As one islander put it, 'What money is in the West, *wontok* is in the islands.' Newspapers often refer to all Solomon Islanders as *wontoks*.

Tuna fish may be very gradually helping to turn all the Pacific island developing nations into *wontoks*, speaking with one tongue among themselves and presenting a united front to all outsiders in defence of their own resources and of their futures.

VI

PERU: FROM THE SIERRAS TO THE SHANTIES

'Neighbours, if we are capable of making do with the money our husbands give us, then we are capable of running this country! Right now men are fighting each other because they don't know how to distribute a country's wealth. I am sure that if a woman were running it she would know how to administer the country's resources.'
A WOMAN OF VILLA EL SALVADOR, Peru

Emerita Castro, a slim, straight-backed woman in her mid-thirties, was watering the few flowers by her front door. It was 6.30 in the morning. She had been up since 5 am, making her family's breakfast, preparing lunch for her husband to take to work, and lunch for her two children to eat when they returned home from school. Then she did the countless, tiny, unplanned things a woman must do to get her family up, dressed and out of the house in the morning. The house in question is a low, two-roomed shanty home with woven-mat walls and a tin roof.

Emerita's house is situated in the poorest section of Villa el Salvador ('Town of the Saviour'), a so-called 'young town' of 300,000 people hugging the southern edge of Lima, the capital of Peru. It is a shantytown, a mass of simple houses, most of them built by those who live in them. Villa el Salvador is also a mass of contradictions.

It is not a maze of twisting, nameless and sunless lanes like most of the shantytowns in Asia and Africa. It is poverty on a grid system; surveyors laid out straight, unpaved streets and right-angled open spaces before the area was occupied by poor families in 1971. Most of the streets remain unpaved today. Spaces were left open for parks, hospitals, markets, businesses and factories; most of these spaces are still empty. Children play football and rubbish accumulates in the squares. The town was laid out in a desert so bare and bleak that it makes the deserts of northern Kenya seem like botanical gardens. Not a tree, not a shrub, grew here before the town came. It rarely rains, so Emerita and her neighbours can manage to live in simply-built houses made with walls of reed matting. The reed walls allow breezes to circulate in the rooms, relieving the stifling heat of the summer which drains the children's energy and makes them lie in the shade for the hottest part of the day. But the same breezes also bring in the desert dust and grit, sweeping it through the walls and depositing it everywhere, even in clothes and food stored in trunks and cupboards. In the winter, damp mists drift in from the cold currents of the nearby Pacific. The houses cannot be heated and the children suffer

Emerita Castro

from coughs, colds and chest infections. Tuberculosis is rampant in Villa el Salvador.

Water is a scarce commodity in the shantytown. Although pipes were installed in the late 1970s, demand has now overwhelmed the system and there are often breakdowns. Many families still rely on delivery from tanker lorries while others can get water from their taps for only forty-five minutes a day; taps are always left open so containers and tanks are filled whenever water is available. (The town has had electricity, more of a luxury for desert living, almost since its beginnings.)

In yet another contradiction of this 'planned' shantytown, Emerita, although living in the poorest section, always has ample water both for her family and for her few flowers. As it grew, Villa el Salvador spread across the landscape down a gently sloping plain. Water pressure is therefore highest in the newest, poorest sections. The flowers give Emerita the feeling that she has some slight control over her bleak environment: 'I have flowers by my front door for two reasons: first, to make the house look nice. It is not much of a house, but they give it some life, the flowers. And also so that they can purify the air – because plants give out oxygen.'

Rabin, Liz and Emerita Castro beside their home

On first appearance Emerita herself is not a striking person. Her face has the flatness of the Quechua Indian people but her nose is sharp, perhaps giving a hint of her character. She wears the clothes of the women of Third World cities around the globe: the cultureless, placeless skirts, blouses and sweaters of cheap, synthetic material on sale in street markets as far apart as Addis Ababa, Bangkok and Rio. Her children – Liz, who is nine, and Rabin, eight – wear similar cultureless clothes at school and at home.

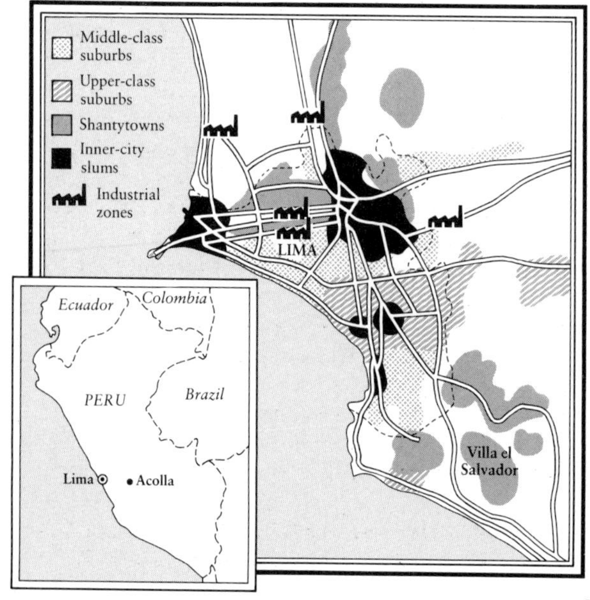

Most of the residents of Villa el Salvador are Indian, although not all are the Quechua people who come from the valleys of the Andes, which form the spine down the middle of Peru. Others are from the forests at the headwaters of the Amazon, or from the lowland coasts. Over 80 per cent of Peru's population of 20 million is either pure Indian or *mestizo* (mixed), with only 15 per cent of the population classed as Caucasian. But those 15 per cent control much of the money and much of the power. The Quechuas are descendants of the sophisticated Inca civilisation, while some of the Caucasians trace their ancestors back to the Spanish *conquistadores* who virtually wiped out the Incas for their gold.

The reasons why Villa el Salvador is populated largely by poor Indians have as much to do with modern world economic forces as with past history. Peru is not an agricultural country; less than 3 per cent of the land is officially classified as arable. Much of it is too steep or too dry to be workable. The government is more concerned about extracting copper, silver, lead, zinc and fishmeal as these natural resources earn foreign currency. So the government has invested little in agriculture, little in the economic activities of the countryside. Any such investment has instead been made by whites of Spanish descent who have amassed huge estates or *haciendas* which produce coffee, maize and wheat for the city markets. Before the land reforms of the late 1960s and early 1970s, about 70 per cent of Peru's agricultural land was tied up in farms of over 2500 acres, and all of these farms were owned by only two-tenths of 1 per cent of the population. The reforms themselves did not go far, transferring according to one estimate no more than 1·5 per cent of the national income, almost all of which went to the 100,000 permanent estate workers who were better off than the peasant farmers in the first place.

With few jobs and little land available in the countryside, poor rural people are forced to come to the cities. Emerita Castro was born and raised in the Andean village of Acolla, ten hours away by train from the capital. Having received her secondary school's degree, and as yet unmarried, she came to Lima alone in 1972 to look for a profession; she wanted to be a nurse, and sought training on a programme offered by the Police Academy. She immediately had to face the realities of Peruvian life.

'I went to the Academy and put my name down; I had all my papers, everything. But I began to look around me. There were a lot of girls applying, all with their mothers, their fathers: a general, a colonel, some captains, being escorted here and there by the Academy staff. To be accepted by the Academy, you needed a lot of influence. I felt completely alone.' She did not obtain a place; the girls who were accepted were mostly of Spanish descent.

After working for a few years in factories, Emerita returned to her village in the Andes where she married a village lad, Marcelo, a skilled mechanic. Together, they decided to make another assault on the capital. Given the visual contrast between the green lushness of Acolla and the gritty bleakness of Villa el Salvador, it is hard to understand why a couple would trade the former for the latter.

'There are more droughts and floods in the highlands than previously,' explained Emerita. This is due largely to the rapid deforestation of the Andes, not for the timber but for the potential farmland once clearing takes place. After forests are felled, water runs off the mountains more rapidly, causing floods in the rainy seasons and droughts in the dry, as little water percolates into the ground to replenish springs and streams. Soil erodes away as well; about a third of Peru's Andean land suffers erosion. The fertile soil of the valleys is now rapidly being covered with cities, whereas before the Spanish estate developers came cities were built only on untilled ground at the valley edges.

There is little productive land left: 'Just little bits; twenty square yards per family, thirty yards; not the large expanses there used to be,' said Emerita. The same is true throughout the other Andean nations of Bolivia, Colombia, Ecuador and Venezuela. These countries have vast unoccupied areas, but the amount of fertile land available to peasants is scarce and most peasant plots are small. Small plots produce small amounts of food. The average Andean dweller eats only about three-quarters of the protein he or she needs and not quite 90 per cent of the calories.

Even for those who are better off and better educated than the very poor and hungry, the countryside offers little hope for the future. There are few non-farming jobs there, according to Emerita's husband Marcelo: 'All a man can do in Acolla is look after the land, work on the land; there is no other sort of work he can do. And there is less land also. Besides, it is a rustic job; young people want progress.'

'Everything is very centralised here in Lima,' argued Emerita. 'The government needs to decentralise the industrial zones, distribute them through each district, province and state so that it is not necessary for everyone to come and live in the capital.'

Looking at the entire developing world, one has the impression that everyone is coming to live in the capitals. The biggest cities are in the South. In 1950, only five of the world's fifteen largest metropolises were in the Third World; by 1980, the South had eight in the top division and by the year 2000, it is expected to have twelve in the top fifteen. In 1950, Africa had one city of over 1 million people – Cairo. In 1980, there were nineteen, and by the year 2000, Africa is expected to have fifty 'million cities', housing almost 20 per cent of all Africans.

It is said – although no one can prove or disprove it – that 1000 new people arrive daily in such cities as Mexico City (population: over 16 million) and Bombay (population: over 8 million). Some are forced out of the countryside by eroded and overcrowded land and lack of jobs; others are attracted into the cities by the hope of jobs, health care and education for their children. For many, as for Emerita and Marcelo, it is a combination of both: slim chances of success in the countryside coupled with great hopes for the future in the unknown capitals.

The invasion from the countryside, added to the high birthrates of families already in the cities, is overwhelming the ability of the city planners and governments to provide affordable land, water, sanitation, transport, building materials and food for the urban poor. It is they, and not the planners, who are determining the directions of city growth by choosing where they put their shanties. Cities such as Bangkok, Bogotá, Bombay, Cairo, Delhi, Lagos and Manila each have more than 1 million people living in illegally developed squatter settlements or shantytowns. Smaller capitals such as Colombo in Sri Lanka, Freetown in Sierra Leone and Nairobi in Kenya do not have such huge numbers, but still have as many as 30 to 75 per cent of their people living in such settlements. Some 4 million of Peru's population live in slums and shantytowns, half of these in Lima.

The problem which Emerita described earlier, of all jobs and opportunities being centralised in one big city, bedevils the development of many Southern nations. When Northern nations were entering their industrial development age, economic activity was generated not just by the capital cities such as London, Paris or New York. Britain could look to Manchester, Sheffield and Birmingham; France to Marseilles and Lyons, and the USA to scores of job-filled manufacturing cities soaking up rural labour.

Compare the growth of London with that of Mexico City. Between 1801 and 1851, the boom years of Britain's industrial revolution, London's population crept slowly up from 1·1 to 2·7 million; by 1901 it was 6·6 million and by 1951, 8·3 million. Since then the population of Greater London has actually declined slightly. By 1950, Mexico City already had a populaton of over 3 million, which has shot up to over 16 million today; and by the year 2000, it could hold 31 million people, according to somewhat improbable UN forecasts. London grew by only 1·5 million over fifty years of rapid industrialisation. Mexico City increased by 13 million over a thirty-year period which in no way could be described as witnessing an industrial revolution.

In so many Third World nations, job opportunities other than in agriculture or very small trading exist almost solely in the capital or in one big city. Although São Paulo contains only about 11 per cent of Brazil's population, most of its high-productivity manufacturing plants are concentrated in the city; Greater Manila has 60 per cent of the Philippines' factories; Nairobi houses almost 70 per cent of Kenya's officially registered factories, but less than 6 per cent of its population.

So, motivated by lack of land and jobs in the countryside and hopes of a better life in the city, Emerita and Marcelo came as a young married couple to Lima in the late 1970s. Marcelo got a series of jobs working as a mechanic. After living in various poor parts of the capital, they took up a vacant plot in Villa el Salvador in 1981. As few people in Peru lack relatives among Lima's 5 million inhabitants, Marcelo called upon his brothers to help him put up the couple's shelter in one day. There is a national bank which gives loans to poor people to help them build their houses. 'But this is a lie,' said Emerita, 'for the first requirement is that your salary must be three times what Marcelo makes.'

In many developing nations, such banks actually transfer money from the poor to the rich. The National Housing Bank of Brazil, for instance, was set up in 1964 and financed by enforced contributions from workers. It quickly became one of the nation's biggest banks. Yet in its first ten years, it devoted less than 10 per cent of its funds to loans to the poor, who make up 80 per cent of Brazil's population. So

D'igir, chairman of the Pastoralists' Association, watches the auction. He has decided his future is in livestock and in the drylands.

Dudley Tapalia (in the red shirt) and his shipmates on the Solomon Pathfinder *amid a school of tuna*

After a school has departed, the deck is a mass of struggling tuna to be packed away in ice

The South Korean lumber company has turned an idyllic beach into an industrial lumber yard (right). *Not all destruction is man-made. A typhoon destroyed property and felled trees throughout the islands in the spring of 1986* (below).

The beginnings of Villa el Salvador, the shantytown in the desert, 1971 (left). *The land is greener near Emerita's home village of Acolla, but it has been divided into plots which are too small to support farmers* (below).

Emerita crochets in her garden, reed matting protecting her from the force of blown sand (right). *A 'popular dining room' in Villa el Salvador, with volunteer mothers helping to prepare the food* (below).

The entrepreneur as opera star: clothing magnate Li Guangming made up for his role in a Sichuan Opera Company presentation (left). *Li's main workshop is back home in the village beside his house* (below).

Niang Yinshu's plastics factory (right). *Niang delivers washing machine and car parts by tricycle* (below).

Queuing in downtown Lima for permission to settle in Villa el Salvador, 1971

the poor, the majority of depositors, were actually financing housing loans for the better off, those who already had the necessary collateral to offer banks.

Villa el Salvador had sprung out of the desert fringes of Lima a decade before the Castros moved to the shantytown. In 1971, Peru was run by the populist military regime of General Velasco Alvarado who was keen on agrarian reform and nationalising the economy. It was high season for 'people's organisations', and one of the main things people were organising was mass land invasions, by which squatters would occupy vacant land in or near the capital. One group of squatters chose to invade prime land near the city centre in early May 1971, when the Inter-American Development Bank was meeting in Lima. The squatters calculated that the police and army would not violently evict them with the international bankers watching. They were so well organised that they had hired architects to mark out and 'zone' the land to be seized. The invasion was successful; by 6 May there were 9000 families on the hillside site. But the squatters had chosen badly; the land they had occupied was *too* good, and powerful elements in Lima wanted to develop it. So the government offered the squatters an alternative site, and General Velasco Alvarado himself made a show of queuing up with the hopeful families to sign up for plots fifteen miles from central Lima in the desert. The site came to be known as Villa el Salvador.

Newly arriving families were greeted by a forest of bare stakes marking plots on sand. Electricity was brought in, haphazardly, in 1972–3. Water pipes, more important in this landscape of little rain, were not installed until 1977–8. But the water supplies and the open storage tanks are dirty, and the children suffer from diarrhoea and worms. The standard of roads is a major issue today; tarmac is only now being laid over a few of the bigger sand streets. There is an area which has been designated an 'industrial park', yet it has no industry and no buildings or roads – only street lamps rising surreally out of the desert sand.

When the Castros moved into a new, poor section of Villa el Salvador in 1981, Emerita found a well-organised community. Being one of the few Third World shantytowns not erected on illegally occupied or illegally subdivided land, it had, from the beginning, a degree of government blessing. This accounted for the installation of electricity and running water. There are shantytowns in Lima which have existed for forty years yet do not have these luxuries. The spirit of organisation behind the original land invasion led to the creation of CUAVES: Comunidad Urbana Autogestionaria de Villa el Salvador (Urban Self-Management Community of Villa el Salvador). CUAVES, which was initially not recognised at any Peruvian government level from national to municipal, established a democratic organisation inside Villa el Salvador: each block of twenty-four families has three representatives in the 'group'; each group of sixteen blocks has eight representatives on the 'communal council'.

In November 1983, Villa el Salvador was finally recognised by the Peruvian government and allowed to establish a city government. Michel Azcueta, an earnest, 39-year-old balding bachelor who had directed CUAVES, was elected mayor, and the first act of his administration was to recognise CUAVES as the sole popular representative of the people, a city council and administration in one. As CUAVES had already built twenty-six of Villa's twenty-eight schools, opened 150 community centres for children between the ages of two to five and planted half a million trees, it made little sense to abandon this proven approach for a more bureaucratic urban administration.

Emerita had become involved with CUAVES and the needs of her neighbours almost as soon as she, Marcelo and the children moved to Villa el Salvador. The local CUAVES people heard that she had a secondary education, rare in the shantytown, and immediately put her to work as a 'promoter'. Her task was to promote health by visiting schools and homes on her block, finding out who was ill, and making sure that they received some form of medical treatment, either from a doctor or a clinic. She does this through the Mothers' Club, and has never been paid for the work.

Life for Emerita in Villa el Salvador has been dramatically different from life in the Andes. In her village she was part of a culture, the culture of the Quechua Indians. She had a firm place in the world; customs that dated back centuries guided her actions and the actions of those around her. Of equal importance was that she and her family knew how to secure the resources of daily life: food grew near by, or was available in local markets; water came from wells and springs; houses were built of earth. Extended families helped sick or poorer members through hard times; there was an informal but tight and efficient system of caring. Not that life was idyllic; it was precarious, but it was predictable.

After choosing to leave her cultural heritage behind for the promises of the city, Emerita and millions like her all over the world have had to build, in some cases almost from scratch, new urban cultures; they have had to find new ways of securing food, water and shelter, of guarding against illness and unemployment. In vibrant economies, jobs, salaries and savings can provide all of these things. Most Northern governments and political systems offer the safety nets of social security systems, subsidised food and subsidised or free health care. But in Lima – and in most Third World capitals – there are not enough jobs to go around. In the 'southern cone' of Lima, where Villa el Salvador lies, there are 1 million people but not a single factory. Peru's social security system is limited, with previous governments spending only about 5 per cent of their funds on health. The state social security system provides separate medical care for those who are formally employed, and through this programme spends 30 per cent of the health budget on 10 per cent of the population.

Emerita's work in promoting health brought about her involvement in the 'popular dining rooms' of the Mothers' Club. Each dining room has a paid cook but most of the work of organising meals and purchasing food is done by women volunteers, who also begged or bargained for the bricks, cement, stoves, pans and architectural advice to set up the dining rooms in the first place. There is roughly one dining room per 360 families. Almost no food is grown in Villa el Salvador, and the small shops in the shanty-town offer food at the usual high prices of small retailers. The women therefore have to travel miles, usually on foot, to the fish or vegetable markets in order to buy fresh food in bulk at reasonable prices. 'Members' of the dining halls pay for their meals, but these meals are cheaper than can be provided by one mother buying food in the small quantities dictated by the lack of refrigerators and the hot climate. Lunch, for instance, costs the equivalent of seven pence (ten cents). If members are sick or experiencing other difficulties, they receive free food.

Almost half of the children in Villa el Salvador are malnourished. It cannot be proved that the dining rooms have raised the children's nutritional standards, but they do provide part of a do-it-yourself social security system. About one in four of the people eating in these dining rooms receives free meals.

'When people have accidents or are ill, they come to the dining rooms to tell us this; they ask for help. Well, that is what we are here for. This is our objective,' said Emerita, who now co-ordinates the women's work in seven such dining halls, any one of which may feed 180 people for breakfast and lunch. Even those who are not members of the dining room can still obtain food and advice.

Emerita knows at first hand the value of this service. Three years earlier, as Marcelo was welding under a car, his torch exploded, burning him badly. He was in hospital for two months with no insurance, no disability payments from his employer and no money coming in to the family. His hands were so badly damaged that he can never again work as a mechanic. Today, a thin and slow-moving man, he can do only unskilled clerical jobs for the minimum wage, the equivalent of £20 ($30) per month, whereas as a mechanic he earned £55 ($83). Emerita and the children ate free at the popular dining rooms while Marcelo was recuperating.

Emerita did not drop her 'volunteer' work and try to find a waged job when Marcelo had his accident. She already spent odd moments during the day crocheting work to sell to shopkeepers. After the accident, she worked longer hours and increased her output of elaborate tablecloths for the wealthy and the tourists. Most of the women in Villa el Salvador sew, knit, crochet or embroider to help make ends meet.

Such work is not very fashionable among writers of books on women and development, some of whom claim that it is sexist to teach women domestic skills such as cooking and sewing when they should be acquiring skills for the 'real world' – farming, metalwork and machining. Yet, despite the theorists' arguments, the sewing and knitting arts fit the needs and life-styles of the women of Villa el Salvador perfectly. They have to walk great distances in order to shop, and they can knit as they walk, or as they wait in a clinic or attend a Mothers' Club meeting. They knit at night while the children are asleep.

'We must work,' Emerita explained, 'but we must do something within the home without having to leave our children. If we went out to work they would grow up badly; they would wander off the right path, so to speak.' Emerita's friend Maria Elena, president of the Women's Federation, described another advantage of such work: 'To work with wood we would need a chisel; we would need a saw to cut it up to make furniture, right? But to make a jumper we just need knitting needles and our energy.'

Such work is part of the 'informal sector', unrecorded in any official statistics, and includes shining shoes, combing rubbish dumps for valuable objects, or turning car tyres into sandals in dark back rooms. Such labour is the major work in the Third World cities. Over half of all of Peru's active labour force works in the informal sector. In Lima there are tens of thousands of street vendors, selling flowers, oil, trinkets, radio batteries and cigarettes; many of them are former professionals who are out of work due to Peru's ten-year economic recession.

However, knitting and most other such informal work brings little reward on the open market. 'We don't know how to value the cost of our energy and we get paid a pittance for it,' said Emerita who, despite this view, spends all of her afternoons at one dining room or another teaching Mothers' Club members how to crochet and knit. Meanwhile, Maria

Emerita teaches the women to knit and crochet. It is work they can do while performing other chores.

Elena and the Women's Federation are teaching women how to market their products and are setting up an association to help women supply their work directly to the shops, rather than through countless middlemen who take all the profits. Today, a woman can work for forty-eight hours on a jumper, then sell it for the equivalent of £1.20 ($1.80) to a buyer, who sells it to a wholesaler, who sells it to a boutique, which sells it to the public for over £10 ($15). 'What we are trying to do is shorten the chain, the chain of exploitation of the mothers,' said Maria Elena.

When she has crocheted several items, Emerita makes the long bus trip to the centre of Lima where she sells her work directly to an Italian woman who owns a boutique, but she receives little more money than if she had sold through middlemen. On a recent visit she sold three months' work for the equivalent of £20 ($30), half of which she would receive only when and if the shop sold the goods. When she visits the buyer she feels 'a little humiliated, a little . . . how can I put it? At first I felt like I was begging, but then I thought, "No, I am selling the work that I do."'

It is perhaps beginning to sound as if Villa el Salvador is a town of only women. For most hours of the day, this is true. The men must rise early in the morning to commute to the centre of Lima, or to other areas where they work or where they hope to find work. Even most of the 68 per cent of the workers in Villa el Salvador officially described as 'unemployed' must leave the shantytown to carry out their informal jobs of collecting scrap or shining shoes. Most of the men leave before sunrise and return long after sunset, six days a week. It takes Marcelo an hour and a half to travel to work on overcrowded and infrequent buses, and it costs him 15 to 20 per cent of his small salary. While the men are at work, the women are left in charge.

According to Mayor Azcueta, the men's absence means that, 'The women are the ones who suffer the lack of water, electricity, education for the children, food, the garbage problem. Without the organised and massive response of the women, I don't think Villa el Salvador would be what it is today. It is they who have planted trees, built many schools, made pavements, administer the dining rooms. The role of the women is crucial.'

It is especially through these chores that the women are forging a new culture; but abandoning the old ways brings both gains and losses. For instance, in Quechua culture a married woman cooks only for her husband. The cooking is both a function and symbol of the union, as basic and sacred as sexual constancy. It is therefore hard for Europeans to imagine the impact which the women's work in public dining rooms has had on their men, on their image of their wives and on their image of themselves. It is equally hard to imagine the liberating effect it has had on the women – overturning in one stroke thousands of years of repression which demanded that women stay close to their home fires. Such a move realises many of the ideals of the Women's Liberation movement of the North.

'There are many husbands who don't want their wives to go out, to participate,' says Emerita. 'The man says, "No, the woman is to take care of my house and my children." They see us as useless, as if women were only made to have a brain just so they can take care of them and cook for them.' Emerita worked hard to convert Marcelo, because she thought it was one thing for him to hear from her about all of the deficiencies of the children's school, but quite another thing to see them for himself. So when he was recuperating from his burns she encouraged him to go to the school and take a look. He became involved – just as he once was with the village council in Acolla – and is now on the executive committee of the Fathers' Association. 'He gets a bit moody sometimes when I neglect the house a little. Then he puts the brakes on and says, "That's enough." But he doesn't stand in my way.'

Emerita is fortunate in her husband. The pressures of shantytown life and of these cultural crises have meant that over 30 per cent of the women in Villa el Salvador have been deserted by their husbands. Abandoned wives with children number significantly among the beneficiaries of the public dining rooms.

Maria Elena of the Women's Federation is a more public, political figure than Emerita. She is younger and more voluble, and her flamboyant 'Afro' crown of tight curls matches her fiery style. In a speech she gave to Emerita's Mothers' Club (a speech which began with a call for stricter attendance at knitting classes), she examined the whole issue of 'women's work' in small and large contexts from family politics to national politics: 'How many of us mothers fill ourselves with children? Why? Because we don't know how to plan our families, just because we are women, just because our grandmothers told us women are good for nothing but washing, cooking and ironing. Do you believe that? Neighbours, if we are capable of making do with the money our husbands give us, then we are capable of running this country! We are as good as men. Right now men are fighting each other because they don't know how to distribute a country's wealth. I am sure that if a woman were running it she would know how to administer the country's resources.'

Resources again. Villa el Salvador has tremendous advantages over most Third World shantytowns. It is a legally constituted municipality. Its people own their own plots and homes. It has well-run people's organisations which have been established for fifteen years, and tremendous human resources in people such as Emerita, Maria Elena and Marcelo. It has electricity and public water supplies – after a fashion. It is adjacent to the nation's capital. It is a familiar name in the literature of self-help urban development, and it receives direct aid from European development agencies and from UNICEF and UNESCO. Yet in spite of all of this, most of its people, who are neither lazy nor stupid, cannot manage to secure the basic resources necessary for decent lives. Children are malnourished. Infant mortality rates are high. Adult health is poor. Most people are unemployed or underemployed; most live below the poverty line.

These facts have persisted under four very different styles of national government. Mayor Azcueta noted

Maria Elena (right) *of the Women's Federation visits a woman deserted by her husband*

that Villa el Salvador came into being during the regime of General Velasco Alvarado, who tried to increase his political power by making appeals to the so-called 'marginal areas': that is, the shantytowns and poor rural regions. Then in 1975 came General Morales Bermúdez who, as economic crisis bit deeper, abandoned these areas and persecuted and imprisoned the popular leaders of the poor. Elections held in 1980 brought a return to democracy in the form of Fernando Belaúnde Terry, who had been overthrown in 1968 by General Velasco Alvarado. Belaúnde set up a programme of *Cooperacion Popular* (the People's Co-operation), giving aid to the marginal areas and the establishment of the people's dining rooms in places such as Villa el Salvador. Further elections in April 1985 brought another change in government, heralding Alan Garcia's centre-left policies, which may yet prove beneficial to the shantytowns. But Mayor Azcueta, speaking during the 15th Anniversary celebrations of Villa el Salvador, made the point that, despite radical changes in national government, there had been no radical changes for the people of his municipality. They remained poor, sick and hungry. 'There have been no housing policies in the different governments, and there have been no changes, neither Velasco, nor Morales, Belaúnde, nor the present. Because they do not reach the bottom of the problem!'

Mayor Azcueta, tieless, shared the podium with Emerita and many other women during the Anniversary celebrations. With simple, improvised questions, such as 'Do you both take each other?' this former student for the priesthood officiated at the marriages of 500 'Salvadorian' couples in a mass outdoor ceremony. The couples were too poor to afford church weddings. Then, with a few friends, Mayor Azcueta spoke of present realities and future hopes. He was proud of the 200 popular dining rooms in Villa el Salvador, but felt that they were only a partial solution to hunger. 'There are no jobs in the whole southern cone of Lima. This is the main problem, and from this problem come all the others, the sicknesses, the malnutrition, the lack of water and electricity. But the first problem is the lack of jobs.'

On the matter of jobs, the mayor was driven to communicate with his national government (only fifteen miles away in Lima) via Austria. He contacted the UN Industrial Development Organisation in Vienna which contacted the Peruvian Ministry of Industry. Azcueta was expecting to sign in the near future a three-way agreement between the municipality of Villa el Salvador, the Peruvian government and the United Nations to develop the bare industrial park in Villa el Salvador to support small enterprises.

However, progress will have to be both quick and substantial. In 1987, some 100,000 new arrivals were expected in Villa el Salvador. 'We are not able to control the immigration toward Lima, and that frightens me,' said Azcueta. 'Up until now we have managed to incorporate new families into the dynamics of Villa, but if 100,000 more do come in a year this whole social experience could be in danger.' Peru's national financial problems – large debts and stagnant markets for its major exports, together with President Garcia's tough line with his nation's creditors and with the International Monetary Fund – mean that the nation will lack for some time sufficient economic growth to soak up the labour force in its rapidly growing shantytowns. 'The economical situation of the country itself, if it continues to get worse, well, our work could fail in every sense of the word,' said Azcueta. Nationally, average wages fell by one-third between 1982 and 1985 – dropping to less than half of what they had been in 1975. It is hard to build a city when its builders become poorer each year.

These threats are not causing Azcueta and Emerita to panic and change their attitude toward aid from abroad. They both feel that such help must be offered only to assist efforts already being made by the people of Villa el Salvador, and that programmes and projects must not be imported wholesale together with the ideals and ambitions of foreigners. 'What is important about help from abroad,' said Azcueta, 'is that it comes to groups, to organised experiences, and that these experiences do not begin with help from abroad. This has been our experience for fifteen years: if something already on-going receives help from abroad, it does not necessarily depend on this foreign aid. But if something starts with aid from abroad, then when this aid leaves, it dies.'

Nor are they willing to be pawns to the ambitions of their own national politicians and parties. Maria Elena warned a dining room full of women: 'The municipal elections are almost here, and when there are elections every party suddenly remembers the women. But we don't know about politics and political parties, so we get used. How many times have we been used? In Villa el Salvador, we have always been women fighting for Villa el Salvador, not women fighting for a particular party.'

'If there is something, it should be shared by everyone. It shouldn't just be for the people with the same politics,' said Emerita. When Maria Elena and Emerita march in a parade with their red sashes and banners, as they did in Villa el Salvador's 15th Anniversary celebrations, their insignias are not of the national political parties, but of the Mothers' Club and the Women's Federation.

The municipality of Villa el Salvador remains in a no-man's-land between the life of the countryside and life in the capital. Outside the national political jostling, the shantytown itself is jostled by tides of national and international economics, but it is not a player in those games. It sometimes seems, especially to outsiders, that the big and growing shantytowns of the Third World are the dumping grounds for the least able, the most passive, for those trapped by historical forces between countryside and city. Having met Emerita, however, it is hard to continue to view the shanty-dwellers as hapless victims, or the even more damning stereotype described by Mayor Azcueta: 'We are sometimes seen as a district full of delinquents or terrorists.'

When Emerita is not thinking as an activist and promoter, she tends to see herself as living in a sort of limbo. 'I have never, never felt "Villa Salvadorena". I have always said I'm from Acolla, because that's where I am from. But my children are Liman. In Lima, life is very hectic. They push advertisements in through your eyes and it makes people want all the things they see in the adverts. Whereas, in the Sierra, although we have television there too, people are different.' On a visit to her mother in Acolla – her mother wearing the broad colourful skirts and the hat of the highland Quechua Indians – Emerita grew nostalgic: 'I miss my land very much. I have lots of memories. I think about the time I used to live here, my adolescence. I compare my life here to the life I live now in Lima; it's very different, very, very different.'

Would she and Marcelo ever go back to Acolla? 'If it were to change. If it were to change, we would go back. If there were a centre to work in, but not as it is at the moment. If there were a place to work, factories and businesses, then yes, maybe ... But now there are only old people staying up there.'

Success in Lima, in the form of wealth and a home in a fashionable suburb, is not her ideal either. Of her visits to those suburbs to sell her crocheted goods, Emerita says: 'At first I feel a little anger inside me, at the vast inequality in the country, in the world. Sometimes I ask myself why they have so much more than I do. But sometimes I ask myself why if these people have so much, why aren't they happy with what they have? Why do they always want more?' She finds the atmosphere of the wealthy suburbs poisonous; there is an air of exclusion and individualism, very different from the comradeship of Villa el Salvador. 'Marcelo and I talk about our economic problems, and sometimes we settle for saying, "Well, we might end up with too much, too." All we want is to make good our little house, so that it's big enough for the children to be happy there.' She does not even want her children to be rich; she wants them to learn to do 'something technical, because that is the main need of Peru – technicians'.

When Emerita was a schoolgirl in the Sierras, her nickname was 'the disciplined one'; her friends looked to her to organise the parties and outings. Today she wants to develop Villa el Salvador 'in a more ordered, organised, disciplined way, so we can all have a common interest in helping our town progress, so we do not have to spend our entire lives low down, in this sand'.

It may prove too big a job for her, and the other mothers, and even for the mayor. But the running of Third World cities has already proved too big a job for those designated or elected to run them. Nor have Northern nations solved the problems of providing affordable housing situated near work for the poor. Neither capitalism nor communism can claim any great success in coping with such issues. But in the South the problem is bigger because the populations are bigger and the resources fewer. The real builders of Third World cities continue to be the urban poor – the majority of people in all Third World cities. They continue to overwhelm the efforts of city and national officials to guide and control city growth. So leaving the job to the poor is not a romantic utopian vision; it is what is already happening. The results have not yet been satisfactory, because the Emeritas and Maria Elenas of this world usually find themselves battling with the authorities, rather than working in partnership with them.

Recent shantytown improvement projects have shown that where the poor are helped – with credit, advice, cheap building materials, water, transport, with building codes that are realistic guidelines rather than inflexible impossibilities – they are by and large equal to the job of city building.

Emerita and her neighbours are already hard at work.

VII

CHINA: GETTING CASH INTO THE COUNTRYSIDE

'Now I am the boss of the factory, the buyer of materials, the accountant, the salesman, the head of personnel. I work; I literally sew the clothes. I sweep the floor each morning. I work in the fields. Besides, I still have to act and sing in the opera. I am overstretching.'
LI GUANGMING, Chinese entrepreneur

Li Guangming represents in one man many of the contradictions that are endemic in modern China. He is at first glance just one of among 800 million typical Chinese peasants. He lives in a traditional brick-walled house with a blue-grey tiled roof in the village in which he was born on the vast, flat plain of wheat and rice surrounding Chengdu, the capital of Sichuan Province in the heart of the nation. A tall grove of bamboos guards his house and its earthen-walled courtyard.

Being forty-six years old, and having started his family well before the government introduced its programme of encouraging couples to have only one child, Li lives with his wife, four sons – aged sixteen to twenty-four – two daughters-in-law, two grandchildren and his mother – a typical Third World extended family of the kind which is rapidly becoming outdated in China. The entire family owns and farms only one and a half acres of ground, again typical of the land shortage in much of rural China. Li rises each day at 5am, sweeps the floor, tidies the house, and often works in the fields with his family.

Li Guangming

Inside Li's house a visitor's expectations of life in rural China are completely overturned. There are two colour television sets, and a video cassette recorder on which Li can watch his beloved Chinese operas and play back his own appearance at a televised banquet in Chengdu held in honour of, and even extolling the virtues of, some of the province's wealthiest entrepreneurs. For Li and his family privately own and run a clothes factory. Last year Li made a profit of 30,000 yuan (about £5400 ($8100); one yuan is worth slightly over eighteen pence (twenty-seven cents)), a fortune in an area where the average peasant makes 400 yuan (£72 ($108)) per year. Although his ventures into private enterprise in this officially socialist nation have fared better than those of most of his neighbours, his efforts and his rewards are typical of the changes sweeping the Chinese countryside.

It is fitting that Li made his fortune in Sichuan, for this province also embodies many of the contradictions of modern-day China. This province, which is larger than France and has almost twice the number of inhabitants, has a population of 100 million people which has risen from 57 million as recently as 1950. If Sichuan were a nation, it would be the world's eighth most populous. It is a poor province; its per capita income ranks it among the lowest five of China's twenty-nine mainland provinces and autonomous regions. Sichuan is more rural than most, with 87 per cent of its people working the land, compared with the national figure of 80 per cent. Sichuan is a crossroads. The industrial city of Chongqing (Chungking), one of China's biggest hinterland cities, where the Yangtze River gathers itself for its long journey down to Shanghai, dominates eastern Sichuan and resembles China's other big industrial metropolises to the north and east. But south and west of Chengdu the mountains and gorges stretch to the Tibetan border and hold many of China's poorer ethnic minorities. Giant pandas live to the west of Chengdu in the mountains, which water the agricultural basin.

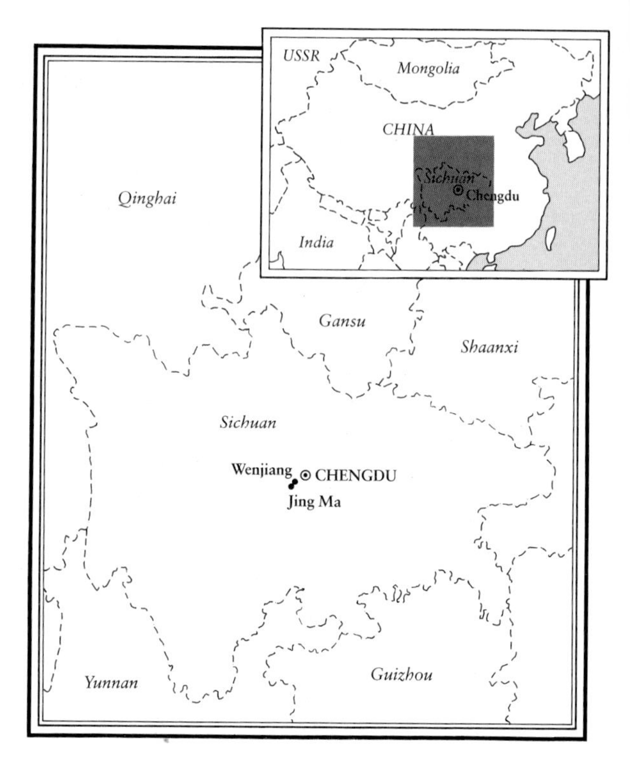

China has always taken its ideologies seriously, from Confucianism and the Legalists of over 2000 years ago to Marxist-Leninism and Maoism more recently. But these ideologies have always been balanced and tempered – sooner or later – with peasant pragmatism. Pragmatism and straightforward mathematics seem to have had a lot to do with China's recent swing away from attempts at a pure 'commune-ism' in the countryside toward a willingness to allow, even to encourage, the rural peasants to follow the pursuit of making money for themselves.

Zhang Yi, deputy director of the Bureau of Township Enterprise Management in the Ministry of Agriculture in Beijing, found himself forced to use very large numbers when explaining the nature of China's human mathematical problem: 'In 1952, China had 262 million acres of arable land. (Only slightly more than one-tenth of China's land is arable.) It had 180 million workers in the countryside, so each worker had a little less than 1·5 acres to farm. By 1985, the total arable land was reduced to 246 million acres, but the workforce had grown to 375 million, so each worker had about six-tenths of an acre.' This is not much land upon which to support a family; but, perversely, there are more workers on the land than China needs in order to provide its food and cash crop requirements. It is a peculiar contradiction: there is not enough land for China's farmers, but there are more farmers than China strictly needs. Zhang calculated that, even disregarding the rapid increase of labour-saving machinery in the countryside, the nation requires only 150 million workers to farm the land and 50 million to raise stock. So the nation has a 'surplus' 175 million workers living and working on the land – a workforce larger than the entire population of Indonesia, the world's fifth most populous nation.

In Peru, and in many other Third World nations where the rural poor are largely left out of economic and political considerations, such 'surplus' people are often abandoned to their own devices and have to become self-reliant. Many, as did Emerita Castro and her husband (see Chapter Six), move to cities; many remain in the countryside, vulnerable to upsets in the delicate balances which control rural life: from falls in commodity prices and unemployment to natural disasters such as flood and drought.

The Chinese authorities did their sums. If these

workers and families moved to cities, the influx would more than double China's current urban population of 200 million, according to Zhang. Creating jobs for the immigrants would cost the government 20,000 yuan (£3600 ($5400)) per person, while another 10,000 yuan (£1800 ($2700)) per head would be needed for social support. Even if only 100 million rural workers headed for the towns, it would still cost 3000 billion yuan (£540 billion ($810 billion)), or about twenty times China's entire money supply, to make provision for them. People must therefore be kept in the countryside, but not as farmers, of which there were already too many. Obviously, people must be provided with incentives to stay there, but Zhang candidly admitted that many of China's past policies had in fact institutionalised poverty in the countryside. He recapitulated briefly:

Just after the revolution there were 2 to 3 million 'peasant artisans' in rural areas – peasants who did a bit more than just grow crops, whether it was making sandals as a sideline or selling meat and vegetables through retail stalls. By the end of the 1950s, after the formation of rural communes, 'commune enterprises' were producing goods worth 10 billion yuan [£1·8 billion ($2·7 billion)] yearly. But as of 1962, no communes or 'production brigades' were allowed any longer to run their own enterprises; the state took over all. This led to such a reduction in small-scale enterprises that by 1963 their yearly production had plummeted to 430 million yuan [£77 million ($116 million)].

During this period, and in fact for three decades following the revolution, the Chinese government was preoccupied with laws and regulations to keep peasants from moving to cities. From time to time there were even enforced exoduses meant to empty the cities into the countryside, a process known as 'sending down'. Such efforts were not particularly successful.

During the period of the Cultural Revolution and the Gang of Four [1966–78], the country's main [urban] industrial output ground to a halt. But in the countryside the situation was very different. With the virtual paralysis of local government, the peasants were largely left to their own devices, and developed small rural industries without government interference. Further, during the 1960s many technicians and industrialists were sent for 're-education' in the countryside. They joined forces with local manual workers in small enterprises, and what today we call 'township or village enterprises' expanded. The plentiful supplies of raw materials in the countryside – cotton, wool, silk, leather, timber, oil seeds and foodstuffs – helped the process. Despite the chaos of the period, production soared, and by 1976 had climbed to almost 28 billion yuan [£5 billion ($7·5 billion)], much higher than in the 1950s.

Today, after a period of government encouragement, there are over 12 million officially registered township enterprises which employ 70 million full-time and 30 million part-time workers producing every-

Li's courtyard and house. The 'factory' is in the back garden.

The 'modern' house of rural Sichuan, made of tiles and breeze-blocks

thing from flowers and mushrooms to rolled steel. The only activities forbidden to these enterprises are the manufacture of cigarettes and armaments. The incentive of profit is thus being used to encourage farmers to cease farming but to stay in the countryside. In 1985, total production from the township enterprises was worth 272·8 billion yuan (£49 billion ($73·5 billion)), ten times its 1976 value; and these enterprises were producing almost 20 per cent of the nation's total production value, which is about the same as China's total production value in 1969, according to Zhang.

Zhang was not offering a personal view of history. In what appears to be a sweeping denouncement of the entire agricultural commune-isation effort, the official *Beijing Review* began an article in June 1986 with this summary: 'After the founding of New China, Chinese farmers were tied to state-owned fields managed by production brigades for growing mainly food grain. Working hard the year round but for a meagre income distributed evenly by the brigade, the farmers, Party members and ordinary people alike, were poor.' The article went on to say that this situation lasted until the late 1970s, when the government reformed the rural economic system and contracted fields out to individuals and groups of farmers under various 'Responsibility Systems'.

In Sichuan Province, Li's own development reflects the ups and downs which the entire nation suffered. His father and a long line of Lis ran meat and vegetable stalls. The family was somewhere between the merchant class and the 'peasant artisans' described by Zhang. As a young man, Li had developed two talents rare in the countryside: he proved himself deft with a needle and talented in singing and acting traditional Chinese opera roles. Thus Li and his young wife were able to do their military service together in the armed forces' Song and Dance Ensemble, entertaining the troops throughout China.

In 1962, the couple returned to Li's ancestral village of Jing Ma, twenty miles south-west of Chengdu, and moved into the house they occupy today. They found times hard: 'We had a large population in this province. We did not have enough land. The whole countryside was very poor.' This was the period during which the government was taking over all rural enterprises, so it did not occur to Li to set up publicly in business as a tailor, even though he and his wife were the only ones in the village with tailoring skills. But he made clothes for his growing family and, as his reputation spread locally, for his neighbours also. He also kept up his singing, appearing with the Sichuan Opera in Chengdu.

He decided to expand his clandestine, word-of-mouth tailoring business. He and his wife collected driftwood from the river and, when they had an entire roomful, they sold the wood to buy their first sewing machine. 'I had to carry that machine all the way home from Chengdu on my back,' remembered Li. He earned enough money from sewing to buy a second machine, and then both machines were stolen. Nothing daunted, he raised two pigs, sold them, and bought another machine.

The Cultural Revolution was the family's lowest ebb: 'The mental pressure was tremendous. We had to do the collective work in the fields. We had to attend political study sessions in the evenings and criticise our neighbours. We had only the government rations, and we had four children to feed.' But the couple continued to make clothes, 'on a private, personal basis', and managed to avoid being criticised – or worse – for capitalist activities by working only at night, often between midnight and dawn. They worked by oil lamp because electricity was rationed and, if Li's electric lights had been seen burning late at night, the authorities would have switched off his power.

But Li was and still is resilient. 'I somehow remember with fondness some of the aspects of even those days. I often had to take the village's pigs to market several hours away. I would walk in front of the pigs and sing. It must have been amusing to see me singing in front of the pigs. I was exercising my voice; I have a good voice.'

When in the early 1980s the assumption of such private enterprise 'responsibilities' became institutionalised, Li was well-placed to take advantage of the change in the wind. He unashamedly admits that, as a Communist Party member, he found it easy to obtain loans from the government and from a bank, especially as the financial backing he required was for an already established business. He built a small workshop adjoining his house in 1983, opened another workshop in the village, bought some sewing machines (costing only about 100 yuan (£18 ($27)) per machine) and invited the local farmers' daughters to come and work for him. Some brought their own machines and were paid on a piecework basis; others used Li's machines and were given monthly wages. Today he runs the two workshops and three other branches of the business which he does not own, but which he supplies with cloth and from which he buys the finished garments. These branches are collectively managed, but in total Li employs fifty people.

He pays his workers 100 yuan (£18 ($27)) a month, a good wage in a region where a typical peasant will only earn 400 yuan (£72 ($108)) in a year, and about the same as a city worker can make. However, when business is slow, Li lays off his employees and they return to work in their fields. 'My children get only 60 per cent of their wages. I deduct and keep the rest because they eat with us and I give them clothes. They keep what they receive as wages. My eldest daughter-in-law came to us as an apprentice, and then she married my son. She is now in charge of all technical matters in the factory.'

Fifty employees far exceeds the government 'guidelines' for a family enterprise, but Li has not had any interference from that quarter. 'The state supports efficiency. The better you run your enterprise, the more support you get from the state. If you abide by the laws, if you pay your taxes, then you will not have problems.'

Li sells his garments throughout western China, regularly making business deals in Tibet and Gansu Province to the north, and recently was planning a sales trip to Canton. Whereas Li travels by bus and train, his sons deliver locally on bicycles, although for longer hauls they use lorries supplied by the local transport department. Li's wife still cuts patterns at home. Much of the business is generated by big orders from wholesalers or retailers. 'But we also design our own clothes. We study and copy the fashions and designs which are popular and produce maybe one hundred to see how they sell. If they sell well, we produce more.' The clothing industry enjoys a sellers' market in modern-day China. On the streets of Chengdu, and even in Jing Ma, one sees the drab blue 'uniforms' of the old style jostling with more and more youths wearing sequinned sweaters, sweatshirts bearing advertising logos, jeans and trainers. Western suits and ties are becoming popular, despite occasional attempts by the government in the early 1980s to eliminate such 'cultural pollution'.

Li attributes his success not simply to his tailoring skills, but to his travels in the entertainment business. The Sichuan Opera gives several performances a year around the country and in Chengdu. Li plays several stock characters, including 'Flower-face', a monk and an old lady's role traditionally played by men, much as the part of 'the Dame' in British pantomime is always reserved for a man. 'People from other places come to the performances. We keep in touch with the outside world. We learn where we can get our materials and where we can sell our products. Through our connections in the entertainment business, we know where there is a demand. We have a Chinese saying: "No friend, no way!"'

Thus Li thinks nothing of travelling hundreds of miles in order to find or close a business deal. His main shop in Chengdu is situated near a bustling new market built in 1985 and near the railway station, a prime site which brings in both dealers and the public. The shop is small and dark inside, having no windows, but Li has installed a good-quality hi-fi system which plays loud pop music to attract customers. He also buys from wholesalers and sells items such as shoes, which he does not produce.

Success has not spoiled Li, a short, balding man with a high forehead and a habitual wide grin. When he dons his wide-brimmed straw hat for outdoor work, little of his face remains visible apart from the grin. A neighbour described him as 'the sort of uncle everyone would like to have'. Perhaps it is his inbuilt personality, or perhaps it is his theatrical experience, but he always appears relaxed and open, neither reticent nor garrulous. When at home, he supervises the huge lunches served round two tables for his family and his workers. When he attended the 1984 banquet for Sichuan's wealthiest men, he wore a traditional costume and sang a snatch of Chinese opera for his fellow magnates and for the benefit of the television cameras.

Like Li's own ascendancy, China's reversal in policy from communes to capitalism under such unabashed slogans as 'Take the lead in getting rich' has been quick and complex; and much of the nation's energies will doubtless be devoted to ironing out those complexities in the future. For instance, there were 52,000 communes in 1980, each consisting of 2600 households on average. These communities have not all simply disappeared; many have opted to stay together and to continue working in the old way. The land is still communally owned, but almost all of it has been contracted out to individual farmers or groups, at first for periods of three years but this was later extended to fifteen. Herders receive a thirty-year tenure on grasslands. The contractors pay tax to the local government, agree to supply a given amount of produce to the state at set prices, and dispose of the remainder of the produce in the most profitable ways possible.

The township enterprises stretch in a tiered system from the largest, the 'township enterprise' in its purest sense, which may involve as many people working together as in the old communes, down to individual efforts. Li runs a family enterprise, while bigger enterprises are carried out by whole villages. Many of the communes sold their machinery – tractors, combines, mills or threshing machines – to the new private enterprises.

The relationships between the privately-run enterprises and the state-run industries can be and usually are extremely complex. The state industries are old-fashioned in that they are burdened by centralised planning and five-year production plans. The enterprises can often offer better products and services more quickly, because they have sprung up solely to take advantage of consumer demand – as demonstrated by Li's quick response to changing fashions. This has led to competition between the enterprises and the state industries, which government officials such as Zhang see as a way of improving the management and products of the state factories. There has also grown up a symbiotic relationship, in that the enterprises may make parts for the state factories, and joint ventures have been undertaken between the two styles of production. Li was considering an offer to take part in a joint venture with a government department elsewhere in Sichuan, establishing a clothing factory for that department.

Niang Yinshu, who lives closer to Chengdu in a village which is half suburb and half countryside, is also involved in a private business which is co-

Niang Yinshu working at one of his plastic extruders

operating with the government. Niang was raised in the local commune with his family, he worked on the farm after school, joined the army and was sent to another part of China. On his return, he farmed his mother's and sister's private plots and ran the commune's restaurant. When the Responsibility System was introduced, Niang's sister-in-law was in charge of a state plastics factory in Chengdu and was perfectly placed to help him set up a small business in that line.

With the money he had saved, and with an agreement from his sister-in-law's factory both to buy his products and sell him raw materials, in 1984 Niang had no trouble in persuading a bank to lend him more money to buy one large and three small plastic extruders – machines which shape hot plastic by forcing it through a die. Six months after beginning his operations, Niang had paid off the bank. He now buys plastic either from the government factory or on the open market, but finds it is cheaper on the open market. His business is registered with the Bureau of Township Enterprises and his premises are regularly inspected by that body, although to the Northern eye the machines may look dangerous, and safety precautions are perhaps not all that they might be. Niang's largest customers are a washing machine factory and a car spare-parts business. This sounds impressive for a cottage industry, but he still makes most of the deliveries himself on his own pedal tricycle.

Niang's austere living quarters, shared with his wife and eight-year-old son, do not have any of the luxuries of Li's more opulent home. There is not even a radio, much less a television. 'I am far too busy running my business to watch TV even if I had it,' he claims.

Niang is unlike most of his fellow rural entrepreneurs in that he is willing to speculate openly about a very touchy subject in 'classless' China: the long-term effects of the entire Responsibility Movement. 'I think the development of these enterprises may lead to the development of different classes in the future,' he admitted. Feudalism – a system based on large numbers of serfs tied to the estates of a small number of rich landowners – dominated the Chinese countryside for some 2200 years in between systems based on slavery and on Communism. Some Chinese political experts fear that the Responsibility System could lead to an oppressive type of capitalism whereby many poor people with small parcels of land or even no land would be forced to sell their labour cheaply to a small number of wealthy entrepreneurs. The hiring of farm labour is now quite common in the countryside. But Niang describes the new system – as he has experienced it – as a very benign form of capitalism.

'I employ eight people, nearly all of them my relatives, though some are very distant cousins, some are from villages far away. They all live in my house and I feed them, as well as paying them well. Many are earning enough to set up an enterprise on their own, and they will do this as soon as they get a bit older and a bit more experienced. Most of these people have their own land elsewhere, which their families till while they work here. But here it is as if we are one large family; if one is sick or has to go home then no one minds and everyone covers for them. I work right alongside my employees, and my wife checks all the raw materials and the end products. Every year I take everyone on a week-long holiday in the countryside, and give a big party for all the employees and suppliers and customers. All the workers are very keen and enjoy working here.'

Indeed the children and young employees are learning capitalism rapidly from their elders, and they are responding to it enthusiastically, in the way that younger generations tend to be attracted to most new trends. Li's third son Li Jih left secondary school at the age of seventeen and helped his father for two years with the expansion of the family business. Then at only nineteen he became wholly independent, opening a new shop in a nearby market town, selling shoes, handbags and, naturally, his family's line of clothes.

Everywhere in and around Chengdu, neighbours of Li and Niang are setting up businesses. A man named Lo started making plastic bags in 1982; his business failed in 1984. He began a haulage business; he now has two passenger buses driven by ex-army drivers. Lo is treading cautiously, as are many of the new entrepreneurs, because he fears that the government may suddenly change policies and move against privatisation, a not unreasonable fear given China's recent history. He has made enough money to be able to make donations to charity and to equip his house with a telephone, a colour television and a huge portable radio, all by the age of thirty-one.

Down the road is farmer Liu, a rice-grower who has converted part of his small plot into a pond in which to raise carp. A few years ago his village ran a business making breeze-blocks, which were heavily in demand by local farmers able to afford new houses and wanting something 'modern'. The new houses, their concrete walls decorated with pink, green and blue tiles, assault the Western eye, but they are status symbols in the countryside. When the Responsibility System was introduced, Liu bought the breeze-block business from the village. His workers collect clinker, the stony residue from burnt coal, from the ovens of local restaurants and deliver it to him by tractor. It is crushed, mixed with limestone and water, patted into moulds with a spade and left to dry. It takes very little capital to start such a business, and indeed such undertakings are the most popular form of private enterprise in the area, in view of the booming market for building materials.

Many of the 'privatised farmers' are doing as well for themselves as the small-scale industrial barons of the rural areas. The Responsibility System allows farmers more scope to grow the crops from which they can make the highest profits – though they must continue to provide the state with set amounts of given crops at state-controlled prices. But the purchase prices of agricultural produce were raised by the government in 1979, another factor which keeps farmers on the land. The profit motive, aided by increased use of artificial fertilisers bought by the farmers but subsidised by the state, has increased

A breeze-block manufacturing plant in rural Sichuan gives the local people modern building materials

agricultural production dramatically, with wheat harvests rising by 12 per cent per year between 1978 and 1985, according to the British journal *The Economist*. China became a net exporter of cotton in 1983, and was able to stop national cotton rationing in early 1984. The availability of sugar, vegetable oils and meat has also increased, and the number of calories in the daily diet of the average Chinese has risen from a subsistence level of 2100 calories in 1977 to a healthy 2700 in 1983. (Most adults need about 2300–2400 calories each day.)

Most of the short-term problems raised by the Responsibility System stem from the very speed with which it has been embraced. Zhang Yi of the Bureau of Township Enterprise Management noted that the main hurdle facing these enterprises is the shortage of both technical and management expertise. (Other basic difficulties facing would-be entrepreneurs include a shortage of investment capital for those seeking to go into higher-risk enterprises and for those with little collateral in the first place, and hindrances imposed by the 'old style' centralised industrial management.) In addition, businesses are springing up more quickly than schools can educate workers; Zhang considers that only about ten out of every 10,000 workers involved in Township Enterprises are properly trained for their work. Three clothing factories in Li's area have gone out of business because, according to Li, 'They were not experts, not professionals, did not have the skills or know-how to run a business, did not know about materials, marketing and information.'

For this reason the government is encouraging technicians and engineers to leave the state-run urban industries and take their skills to the private enterprises in rural areas. Technicians seem willing to do so, because they are offered government-subsidised housing and they have a better chance of receiving advanced education and research grants if they commit themselves to a period in the countryside. They also have the opportunity to earn at least a part of the profits being generated. Government training schemes have been starting up in rural areas, and the government has begun a programme of nine years' compulsory education for all, which is to be introduced slowly over several years.

There are other, perhaps longer-term and less tractable problems which critics outside China are more willing to discuss than those inside the country. When farmers leapt at the chance of taking responsibility for their own land, the land of the communes was divided into tiny strips, some too narrow for easy access of even a handcart. As long as the farmers are using hoes, they can manage, but mechanisation becomes impossible and there are constant feuds over boundaries as well as accusations of theft and cheating. The reallocation of the collectives was not always rational or orderly; many buildings and machines were simply ransacked. The new-found wealth has meant more and bigger houses, often built on fertile fields, further diminishing the amount of farmland.

The race for profits may degrade the environment. First, many of the new rural industries depend on agricultural raw materials. The need to produce these locally, near the industry, may encourage production on marginal land, and indeed peasants in some areas have begun to cultivate steep slopes previously left untilled. Trees are being cut down in vast numbers to provide building materials, and even telephone poles are being surreptitiously felled for the same purpose near Beijing. The World Bank has openly voiced its worry that 'limited scope for increasing domestic production of raw materials could restrain development of light industry in the 1980s'. Second, the increased use of nitrate fertilisers and the effluents from small industries is polluting waterways in some areas, and an official from the east-coast province of Zhejiang noted: 'The foul smell from polluted rivers and ponds is now a landmark of some rural towns.'

Zhang admitted to the existence of these environmental problems, but said that there was little pollution generated by the Township Enterprises as there were few heavy industries among them and strict regulations were being imposed. (Whether some 12 million small industries can be effectively regulated remains open to question.) Zhang maintained that the major industrial pollution challenge was persuading the old state-run industries to clean up their operations.

China still has to grapple with both the practical and the ideological issues raised by encouraging free enterprise in what is officially a socialist society. The ideological approach remains vague. Chinese leader

Deng Xiaoping has admitted in key speeches that certain 'negative phenomena' had arisen during the economic reforms, but insisted that the reforms served China's socialist development. He upheld the need to study Marxism – for guidance rather than as a dogma – but noted that there was also a need to develop professional skills. The government's guidance to local Communist Party officials is becoming rather confusing, stressing that there is no contradiction between 'adherence to' and 'development of' Marxism, as Marxism must develop in order to avoid the production of dogmatists unable to cope in the modern world. (Deng was born in Sichuan, and Chinese Premier Zhao Ziyang was Communist Party leader there from 1975 to 1980, two facts which may help explain why this agricultural province in China's heartland far from the more industrial east made such quick progress in its grasp of the Responsibility System.)

Critics outside China argue that the practical problems of the Responsibility System are more daunting than the ideological ones. William Hinton, a close observer of China, compared what is happening now with what took place in England during the time of the 'enclosures' in the fifteenth and sixteenth centuries when peasants were forced off what was once common land and had to sell their labour as the only resource available to them. Some Chinese peasants are growing wealthy, as they are exhorted to do, but the majority are being left behind. The system has denied land to many of those who did not contract for it early enough: 30 per cent of the peasantry, half of whom had failed to find any work by January 1985, according to Hinton, whose figures remain highly controversial. Just as the most resourceful peasants will do well, so might those areas with the most resources. The regions with the most fertile land and most natural resources could surge ahead, while the backward areas remain backward. China could become a mosaic of rich and poor regions, with a high rate of internal migration between the extremes.

Li Guangming again offers in his own person an example of the contradictions facing the nation. Although he has done what his government asked of him in growing rich by tailoring, he said, 'My main difficulty is people's jealousy and resentment of our higher standard of living. They are a minority, but they do not have the financial resources, or the necessary skills, to set up a small factory or engage in other sidelines. That jealousy cannot be removed until everybody's living standard is raised.'

Li has even gone so far as to request that local government authorities buy his business and keep him on as manager. They are unlikely to do this, as it would fly in the face of the entire movement toward privatisation. 'I would only be in charge of production. Now I am the boss of the factory, the buyer of materials, the accountant, the salesman, the head of personnel. I work; I literally sew the clothes. I sweep the floor each morning. I work in the fields. Besides, I still have to act and sing in the opera. I am overstretching.' It is an understatement from a man who rises each day at 5 am and then tries to do all of his other chores and get to bed by 1 am the next morning: 'Often I cannot even get to bed by then!'

'But they want me to continue to do the things I am doing. They think I have made a lot of people rich, and they don't want the trouble, the responsibilities, that I have.'

It may be that China itself is 'overstretching', though it is hard to know how a nation of more than 1 billion people could avoid doing so. However, throughout its history of thousands of years, with its different empires, kingdoms and governments, China has often had a bureaucracy able to reach far into the countryside and to order the activities of its common people, an ability surprising in such a vast and heterogeneous land. It is true that the results have not always been for the best. But this ability puts China far ahead of the governments of most developing countries, the majority of which have little contact with their rural poor. It gives China at least the potential for sustaining the development of its Responsibility System, for making adjustments as challenges arise.

After all, China was the first developing country seriously to do its sums by taking into account not only the cost of urban migration, but the likely population growth and the amounts of natural resources available for future generations. These calculations shocked China into introducing its controversial programme encouraging couples to have only one child. Nations such as Bangladesh, Indonesia, Nepal, Pakistan and many African countries may find similar programmes necessary; but they may find the courage to embark upon them only when it is too late for such programmes to stave off hunger and social disruption.

China's attempts to save its cities from an onslaught of farmers, and to introduce opportunities in the poorer countryside, also remain controversial. History will judge both the success of the Responsibility System and the government's ability to change and adjust it as the need arises. Again, other Southern nations may have to try similar measures, and many may begin only after their cities have become little more than expanses of vast, jobless shantytowns spreading outward from a few towering office blocks.

VIII

BRITISH ORGANIC FARMING: ONLY MUCK AND MYSTERY?

'We are just beginning to find all of the problems that these chemicals are causing. You use a drug to get certain kicks, but after a while you have to step up the amount of drugs you are taking to get the same kicks, the same results. Before too long you are addicted to the thing.'

EDDIE FEWINGS, British organic farmer

Eddie Fewings does not look like anybody's idea of an organic vegetable farmer. In fact he does not look like a farmer at all.

A small, wiry man in his mid-fifties, Fewings gives an impression of frantic hyperactivity rather than of the slow, placid approach to life normally associated with men of the soil. Expressions of delight, anger, suspicion and concentration chase one another across his face. His bright blue eyes radiate energy; his mouth narrows grimly when he concentrates, and when he has made a telling point he smiles broadly, revealing uneven teeth. He habitually talks with such energy and fervour that his ears wiggle up and down. All in all, he looks more like a car salesman than a purveyor of wholesome organic produce. He is, by his own admission, 'more of an entrepreneur than a farmer'.

Eddie Fewings

Neither does the ninety-acre farm which Eddie manages offer an idyllic vision of the unspoiled English countryside. Like most of the farms on the sandy loams south of Reading, the land is flat, there are few trees and none of the flower-filled hedgerows which break up vistas and add charm to the agricultural landscape in other parts of Britain. The outbuildings are scruffy and the plastic-covered greenhouses in which seedlings are raised are positive eyesores. Outside the Fewings' bungalow is a pond, full of flag irises and frogs which provide at least the background noise of a natural setting, and add the only hint of character to this typical urban-fringe farm.

Eddie Fewings came to Springhalls Farm by a long, circuitous route which began, improbably enough, in the East End of London where he was born. After a period in the army in the 1950s, he spent eighteen months working on a greengrocer's stall in Spitalfields Market. He left this to go and help his father to develop a market garden in Hockley, Essex. They built up the trade in a delicatessen in Woodford. The business was very successful: 'It was known as the local Fortnum and Mason's,' recalls Eddie.

In 1974 they sold the shop and Eddie took off round the world with his wife Norma and their three children. They were away for three years, during which time Eddie made 'some nice little deals in antiques' while he was in New Zealand. In 1977, he returned to England and went to develop a market garden in St Germans in Cornwall. A short spell as a horticultural adviser in Saudi Arabia was followed by a year establishing a garden centre in London's East End. In August 1983, Eddie began work at Springhalls Farm, which is owned by a wealthy Syrian businessman who has invested in British agriculture to make a profit.

Eddie immediately decided – with the encouragement of the landowner, who is opposed to the use of agricultural chemicals – that the land should be farmed organically. 'We didn't have much choice,' said Eddie. 'The soil was in such terrible condition after years of chemical treatment. Had we carried on working it conventionally I don't think it would have produced food at all in five or six years' time. We had to get the soil back into good condition.'

High-tech chemical farming can strip humus and organic matter from soil. Organic farming rules out the use of pesticides, herbicides and artificial fertilisers, favouring only stipulated animal and composted manures and natural additives. These natural fertilisers put organic matter back into the ground and make soil better able to retain moisture in dry periods. In addition, they help drainage and aeration when there is excess water. By absorbing surplus rainfall, organic matter also slows water-induced erosion.

Eddie thus placed himself firmly with a small minority of British farmers – about 1 per cent by some estimates – who have decided to forsake the use of chemicals and to 'go organic', a move which is more easy in the contemplation than it is to carry out. A few years ago men such as Eddie were ridiculed by the agricultural establishment. They were seen as eccentric and anti-progressive, and they were accused of practising 'muck and mystery' farming. It was generally believed that there was no real market for organic produce, apart from a few health food shops, and that the movement towards organic farming was a fad that would pass.

However, quite the reverse has happened. Today, there are over 400 farmers in Britain who have been permitted to use the Soil Association's symbol on their produce in shops as the consumer's guarantee that the food is chemical free. About twenty new farmers apply to the Association each week to have their land inspected so that they too can use the symbol, and two-thirds of applicants are successful first time. The Association is a private charity which, by general agreement rather than law, lays down standards for organic farming in Britain. There is still no legally acceptable definition of 'organic farming' or 'organic produce'.

Until recently, organic farmers either sold their produce to the conventional market (and received the same price for it as conventional producers) or sold it through the few select shops which were prepared to take the organic produce. Then in 1983, a group of organic farmers from Wales decided that organic food should be properly marketed, and they set up a wholesaling operation which they called Organic Farm Foods. The business began modestly, starting with just a chair, a table and a tiny corner of a warehouse in Clapham, South London. Within a year it had taken over the whole warehouse, and by 1986 it had undergone such expansion that it had to move to new premises near Sunbury. During the nine months up to May 1986, turnover rose fifteen-fold, from a modest ten tons of vegetables a week to 150 tons. Sales of organic vegetables were estimated to be worth about £1 million ($1·5 million) in 1986, but there were predictions that this figure could soar to £34 million ($51 million) in just a few years.

British organic farmers cannot keep up with demand. In some months as much as 95 per cent of the produce in this Sunbury warehouse may have been imported from nations such as France, Spain, Holland, Belgium and Israel, according to Organic

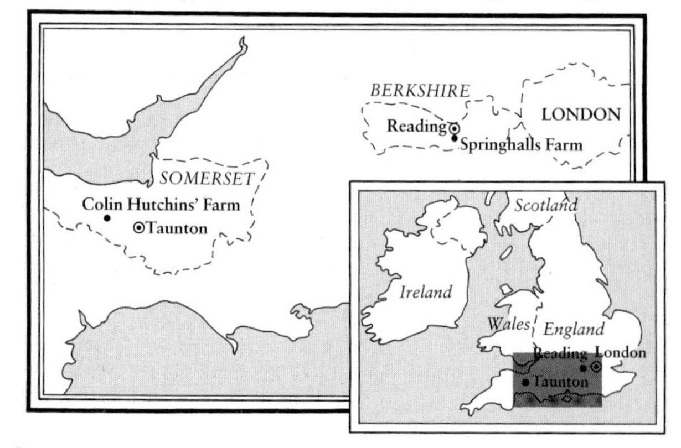

Farm Foods manager and buyer Clive Green. An estimated 65 per cent of all the organic food sold in Britain is imported. Naturally, it is more difficult for wholesalers to check the organic pedigree of this imported produce. But Green insists that organic monitoring organisations on the European mainland are even stricter than the British Soil Association. If for some reason they are forced to relax standards and permit the spraying of chemical pesticides, as happened in 1985 and 1986 when an invasion of fruit flies threatened the apple crops of French farmers, Green learns of it and does not buy the affected produce. 'The market is booming. Occasionally our farmers may over-produce a few lines, when they all grow the same thing, but that is just a matter of a lack of communication,' Green maintained.

The changed attitudes of the big high-street supermarket chains provide the best evidence that organic farming is here to stay. The Safeway chain sells organic produce in most of its stores. According to Dave Cornish, the Safeway buyer, they have only one problem with organic food: they cannot get enough of it. This large and growing demand was the main reason why Eddie Fewings was confident that he could make money, once the farm was fully organic. 'The market's there,' he said; 'it's just a question of getting the vegetables into the shops at the right time.'

Those who see little future for organic farming (and they are growing fewer; even the Ministry of Agriculture is training some of its advisers to give advice on organic methods) generally claim that yields are much lower when chemicals are not used. But so few studies have been made that it is impossible to prove this. Ideally, any study should span two or three rotation cycles (i.e., ten to fifteen years), yet the most comprehensive survey carried out in Britain covered only one year. On average, organic farms were found to carry only about 80 per cent of the livestock per acre which conventional farms support, while cereal yields were approximately 90 per cent of those received from standard farming methods. However, the range of performance was enormous. Some organic farmers had higher, and others much lower, yields than conventional farmers.

Eddie Fewings maintained that his yields are either as good as those of chemical farmers, or slightly less. His main problem has been the transition period, the time it takes to change fields over from a diet of chemicals to one of muck and sunshine. In the first couple of years without chemicals, yields tend to drop considerably. Furthermore, during the transition period, the farmer cannot market his food as 'organic'. Only once he has been awarded the Soil Association symbol can he receive the higher prices which the label guaranteeing that his produce is organic commands. 'Last year we only had twenty-five acres classified as organic,' says Eddie. 'We added another thirty-two acres this year. So we haven't been organic for long; and we're still learning all the time.'

Weaning most of his fields off chemicals has taken Eddie over two years, which is about the minimum time it takes for the changeover; the process can last fifteen years, as farmers struggle not only to eliminate chemical residues but to build up a 'natural' fertility using manure, compost, bone meal and rotations of clover to add nitrogen to the soil. The change is never easy, and the transition period inevitably brings with it financial losses.

It was these losses which made the board meeting in May 1986 at Springhalls Farm a somewhat tense affair. Gill Hall, the representative of the farm's owner, and herself one of the directors of the farm, is demure and soft-spoken but a tough negotiator. She listened intently as Eddie Fewings and his son Jason described the crop losses caused by the unfavourable spring weather. She proved understanding about the farm's problems.

'How are you going to achieve the forecast for vegetable sales in June?' she asked Eddie.

'We're not,' he replied. He outlined which crops they had in the ground and which would be harvested soon. He had already planted ten acres of lettuces, and some mangetouts would be going in soon.

'Is there anything we could sell off now?' asked Gill.

Jason suggested they could raise £900 ($1350) by getting rid of the flock of thirty sheep. There were also the chickens and the four chicken houses, which would bring in £1000 ($1500) – hardly an amount of money likely to impress a wealthy foreign owner.

'We've reached the limit on our overdraft,' explained Gill. 'And we've got an ever-increasing

Eddie, his son Jason, and Springhalls' financial director Gill Hall at a meeting about the farm's future

number of creditors. But now is an expensive time of year, what with the high labour costs and so forth. If we're going to continue in business we've got to inject some more capital into the farm. How much do you need?'

'Ten thousand pounds,' suggested Eddie. 'I think we've got to give it till the end of September. That's when our main season ends. If we haven't turned things round by then – if we're not making a profit – then we should start looking at alternative ways of making a living. Perhaps even sell the farm.'

'Yes, we must give it one more chance,' said Gill. 'It would be nice to break even come September; nicer still to make a profit.'

Eddie and his son agreed; throughout the following summer they thought about little else.

That a vegetable grower in southern England should find himself in trouble in 1986 was hardly surprising. It had been a long, hard winter, the second coldest of the century. Also, since Spain and Portugal had joined the Common Market, thus becoming party to the Common Agricultural Policy (CAP), many small market gardeners growing non-organic vegetables in Britain had been pushed out of business, unable to compete with the cheaper produce coming from the Iberian Peninsula. Cereal farmers (and for that matter sheep and beef farmers) receive a guaranteed minimum price for their produce. If there is a glut, the excess is taken into 'intervention storage'; hence the butter mountains, milk lakes and so forth. In contrast, the vegetable grower is at the mercy of the free – and often fickle – market. Going organic, despite the initial difficulties, gives farmers higher prices for their produce and, much more important, enables them to enjoy a seller's market – a rare occurrence for Northern farmers in this age of Northern surpluses.

In order to understand why an increasing number of farmers are forsaking agricultural chemicals, it is necessary to know how it was that farmers came to rely on them so heavily in the first place. In fact, the use of chemicals for farming is a relatively recent phenomenon. By 1939, British agriculture was in a state of severe depression. Over two-thirds of the nation's food was imported from the Empire, and little money was to be made from farming at home. The Second World War and the German U-boat blockade meant that imports could no longer arrive in the country, and Britain was forced to 'dig for victory' by going back to the land. The introduction of rationing jolted the government out of its apathy, and since then agricultural policy (both in Britain and Europe) has been geared towards increasing the productivity of the land. Every scientific and technological innovation available has been used in the effort to produce more and more food, and modern methods of farming are virtually unrecognisable from those which prevailed fifty years ago.

Before artificial fertilisers became widely and cheaply available, farmers had to maintain soil fertility by combining the rearing of stock with crop production. Crops stripped the soil of nitrogen and other elements; animal manures replaced them, and the planting of crops such as clover helped to 'fix' atmospheric nitrogen into the soil. Farmers controlled pests and diseases through old techniques

A farm worker with Eddie's sheep, which provided manure for his fields

rather than by chemical prophylaxis: crop rotation was their main weapon.

There is no doubt that the staggering rise in farm productivity since the Second World War owes much to the use of chemicals. (In 1983, Britain became a net importer of manufactured goods and a net exporter of cereals for the first time since the industrial revolution.) Since the mid-1950s average wheat yields have risen from one ton per acre to three tons. During the 1970s, yields for wheat improved at an annual rate of 4·3 per cent, those for sugar beet at a rate of 7·8 per cent and those for oil-seed rape increased by 5·5 per cent. These increases are spectacular, although they seem modest when taken in the context of the increased use of artificial fertilisers. Over the last thirty years the consumption of nitrogen fertiliser in Britain has risen nearly tenfold. There has been a considerable 'overkill', as figures from Germany testify: between 1955 and 1965, a doubling in the number of nitrogen applications resulted in a 25 per cent increase in wheat yields, whereas between 1971 and 1980 a 425 per cent increase in fertiliser consumption was accompanied by a relatively meagre 15 per cent rise in yields.

One problem with nitrogen fertilisers is that they are made largely from petrochemicals. As one ecologist has pointed out, we no longer eat potatoes made from solar energy but from oil and gas, which are non-renewable sources. The main use of energy on the farm is as direct fuel and power (29–34 per cent), followed by fertilisers (22–34 per cent) and imported animal feed (13–15 per cent). Sooner or later – as petrochemical feedstock supplies become scarce, and prices again rise – farmers will be forced to stop using such massive quantities of fertilisers. Conventional farming is simply unsustainable.

Of more immediate concern than dwindling oil and gas supplies, however, are the threats which nitrates pose to the health of both consumers and the environment. Only about half of the nitrogen which a farmer applies to his fields is taken up by the crops: the remainder is leached away into the groundwater or 'denitrified' by bacteria, a process which sends it back into the atmosphere as gaseous nitrogen. In some cases, as much as 80 per cent of the nitrogen applied to grasslands turns into waste. Typically, nitrates move down through the soil at the rate of about three feet a year, which means that nitrate levels in drinking water drawn from the aquifer sixty-five feet below the ground may reflect the fertiliser usage of twenty years ago.

Already nitrate levels in some parts of Britain exceed the maximum acceptable level for drinking water laid down by the EEC. Excessive nitrate in bodies of water encourages the growth of bacteria and algae which gradually deprive the water of oxygen and turn ponds and streams into fetid soups devoid of higher forms of life. When humans drink nitrate-tainted water, the nitrates are converted to nitrites in the body. Nitrates have caused cancers in laboratory animals, and are suspected of being associated with stomach cancer in humans, although any link has yet to be proved. They are more firmly linked with the very rare disease methaemoglobinaemia, which is particularly dangerous in babies because it starves their blood of oxygen. (Nitrogen has another, though far less serious, side-effect; it shortens the shelf-life of many vegetables. An organic lettuce lasts on the shelf four times longer than does a lettuce grown with large doses of nitrogen.)

Since Eddie Fewings took over at Springhalls Farm and substituted manure for chemicals, the condition of the soil has improved. However, Springhalls is not a perfect example of an organic enterprise; it is not a largely self-contained, closed system because Eddie has to buy his manure from neighbouring farmers. When he had the sheep, their manure helped, but they provided nowhere near the quantities he required to keep the land healthy.

From the purist point of view, Colin Hutchins' farm in Somerset is about as close to the ideal of a self-sustaining organic operation as can be had. Eddie visited Colin's 320-acre mixed farm near Wiveliscombe on the day when it was scrutinised by an inspector from the Soil Association. Colin is in many ways the opposite of Eddie. Colin is big, well-built, with gingery hair and a strong, reposed face. He talks in a pure Somerset accent, while Eddie still has echoes of the East End. Colin's farm is the sort that the English Tourist Board would want to use in its brochures: hilly and well-forested, dotted with occasional clumps of orchids, the hay meadows rich in herbs and the thick hedgerows splashed with the purple of foxgloves. The outbuildings are rustic looking; the imposing farmhouse is Georgian. While Eddie's 'stock' is limited to a few hens, Colin keeps a herd of white Charolais cows, as well as other beef breeds and a flock of sheep. He has no need to buy manure.

Unlike the many new organic farmers who formerly used chemicals and then decided to give them up, Colin has never used them, except in the most half-hearted way many years ago. He has been farming organically for years (and thus easily succeeded in his application for the Soil Association symbol). However, it was only when he got in touch with Eddie that Colin considered marketing his produce as organic. During the spring he had planted forty acres of vegetables, and he was hoping to receive a premium of 10–20 per cent for the produce by selling it to several wholesalers and retailers using Eddie's entrepreneurial and marketing skills. In order to keep the fertility of his land high, Colin relies on spreading lime and slag and the manure of his animals. He also ensures that his grass leys – land put under grass to rest and replenish it – always contain a large amount of clover.

'I tend to follow two straw crops with a root crop,' he explained to the inspector, who was as interested in farming techniques as in chemical residues. 'Then I go back to grass and clover, and the clover gets the natural nitrogen back into the soil.' Both Colin and Eddie also rotate leguminous crops such as beans and peas. Colin pointed out that his crop rotations not only help to maintain soil fertility, but they also keep pest and disease problems down by denying these scourges the time they often require to take hold of a crop planted year after year. Before any problem has arisen, a new crop has been sown.

There are other, equally 'natural' ways to protect crops from pests. 'When I was a chemical farmer,' recalled Eddie, surveying his own fields, 'my lettuces would get four or five chemical treatments a year – herbicides, fungicides, pesticides, the lot. But now I let nature do the work for me.' As he walked round the rows of lettuces and beans at Springhalls Farm in early July it was difficult to find a plant without at least one ladybird on it. 'This is the little chap that does the work for me,' explained Eddie. 'He keeps blackfly and aphids down to a manageable level; and to me, that's what organic farming is all about.'

Colin and Eddie were bombarded by the opposite approach to pest and weed control when they visited Britain's 1986 Royal Agricultural Show at Stoneleigh. The chemical lobby in the Western world is enormously powerful, and nowhere can one have a better glimpse of its sophisticated approach to marketing

Eddie and Colin Hutchins discuss the future of organic farming over a pint of less than organic lager

than at such big farming shows. Colin (who was exhibiting some of his cattle) and Eddie were both mildly bemused by the elaborate displays mounted by ICI, Norsk Hydro, Shell and other major manufacturers of fertilisers and pesticides. The range of available chemicals is staggering, and their tradenames descriptive of their purpose: Fusilade, Ambush and Assassin. While many farmers use ICI's Fusilade and Assassin to kill wild oats, weeds and black grass, Colin and Eddie use skill. 'I tend to work the ground in early spring and let it lie,' explained Colin. 'The weeds germinate, then I kill them by working over the field again before sowing – that way I avoid getting any serious problems with weeds.'

The agrochemical companies are worried by the growing trend toward organic farming and organic eating, as well as by the rising public concern over environmental damage caused by chemicals. In an ICI advertising campaign one newspaper advertisement showed a small loaf of bread priced at forty pence (sixty cents) and claimed that its price would be £1.60 ($2.40) were it not for the use of fertilisers (somewhat unlikely, as only 19 per cent of the energy required to produce a loaf of white bread goes into growing the wheat, and over 11 per cent of that is in the fertilisers). Another advertisement, aimed at the very heart of British cultural and sporting sensibilities, showed a farmer on a tractor ploughing up a cricket pitch during a match to plant grain – suggesting that such a scene might be expected if the nation were trying to feed itself without using fertilisers. Ironically, the ICI campaign coincided with indications from the Ministry of Agriculture that farmland might have to be taken out of production in order to cut surpluses. An ICI brochure explained to farmers that the advertising campaign was being run because 'the emphasis now placed on "healthy eating", not to mention surpluses of one kind or another, all contribute to farming's bad press. On the strength of such arguments, there's a very real danger that you, the farmer, may unfairly and for the wrong reasons be restricted in the enterprises you farm, the sizes of those enterprises and the way you use farming inputs.'

As for pesticides, the 1983 British sales of these chemicals amounted to £540 million ($810 million); half of this figure was derived from sales abroad, some of which were for chemicals whose use is banned in Britain. About 99 per cent of all Britain's vegetables and cereals are sprayed with one or more pesticides and over one-tenth of all the winter wheat in England and Wales is treated with nine or more different chemicals. No pesticide kills just the target species; most kill many other creatures in the process, including beneficial predatory insects such as the ladybirds which Eddie Fewings values so highly. (However, the chemical companies say that they are making their pesticides more 'target specific'.) Friends of the Earth have identified some 170 pesticide products in use in Britain which contain chemicals about which there is serious cause for concern. Weedkillers can also be hazardous to humans; the herbicide 2,4,5-T contains impurities of dioxin, a chemical which causes death, liver damage and birth defects in laboratory animals but which in humans has been linked only to skin rashes. Nevertheless, in 1984 seven US chemical companies settled out of court with thousands of US veterans who claimed injury from exposure to the dioxin in the defoliant 'Agent Orange' sprayed on jungles during the Vietnam War.

Obviously, those most at risk from the use of these chemicals are the farm workers whose task it is to spray them on crops; but many of these substances make their way into the stomachs of the consumers, and it is this fact more than any other which has stimulated the demand for organic food. When public analysts recently carried out spot-checks on 178 vegetable and cereal samples, they found pesticide residues on sixty-one of them. DDT was found on blackcurrants and lettuces and aldrin was discovered on mushrooms. Many of the pesticides applied to wheat may be found in their original form in a loaf of bread. There are no legal controls to limit the levels of pesticide residue that may contaminate food. The vetting of new pesticides in Britain is carried out under an agreement between the agrochemical industry and the Ministry of Agriculture, but neither the general public nor farmers have any way of finding out how the products were tested or what criteria were used in making decisions to allow them on to the market.

A further problem with pesticides is that pest species develop a resistance to them. By 1983, there were more than thirty pesticides being used in Britain with

Eddie's fields look untidy but he does not mind weeds which do not interfere with the main crop

resistance problems. Worldwide, over 430 species had developed a resistance to chemicals designed to kill them. In some ways, this phenomenon suits the chemical industry: there is always a market for new types of pesticides. However, in many parts of the world, pesticide resistance has led to massive problems. For example, in the 1960s, over 700,000 acres of cotton had to be abandoned in northern Mexico after the cotton pests developed a resistance to all available pesticides. This immunity had occurred precisely because farmers had used too much poison, too often, over the previous decade.

Eddie Fewings summed up the problems of both fertilisers and pesticides neatly, if unscientifically: 'We are just beginning to find all of the problems that these chemicals are causing. You use a drug to get certain kicks, but after a while you have to step up the amount of drugs you are taking to get the same kicks, the same results. Before too long you are addicted to the thing.'

Others worry about the possible effects of these 'drugs', the pesticides and herbicides, on humans, who have only over the past few decades been consuming large quantities of them. 'There are still rather a lot of us, with no declared objectives and no particular desire to discredit capitalism, who would like to know why we and our children have, unwittingly, become the first generations to be fed a diet of chemicals concealed in the basic foods we eat,' said James Erlichman in his book, *Gluttons for Punishment*. He called himself and his contemporaries the 'Guinea-Pig Generation'.

Many farmers have turned against the use of chemicals and gone organic for a number of reasons.

(One farmer made the change when he found that the flood of pesticides he was using had either killed off or driven off the partridges from his farm.) Eddie went organic to improve his soil and to increase his profits by entering an apparently insatiable market. Other farmers also stress that organic farming can save money, and thus increase profits, by reducing the cost and amounts of 'inputs' such as chemicals and fuel. A survey found that 28 per cent of newly organic farmers had made the change because of lower costs and lower capital requirements. Another report summed up motivation in broader terms: 'A major factor has been the mounting unease of agricultural researchers, advisers and farmers with the direction and pace of an agrochemically dominated agriculture and the problems it is creating ... Soil health; good husbandry and stewardship of the land; concern for human and animal health; the fear of agrochemicals, the desire to improve food quality and an active concern for the environment being the major reasons.'

In the USA, where some farming regions suffer unsustainably high rates of erosion, farmers with no fear at all of chemicals are moving part of the way towards organic farming simply to conserve both their soil and their cash. Fuel bills have risen; interest rates on loans have been high, and farmers are spending over \$8·5 billion (£5·7 billion) per year on fertilisers. Thus there has been a trend toward 'conservation tillage', by which method farmers leave mulch on the land and drill through it to plant seeds. Less fuel is consumed, as there is no heavy ploughing; less fertiliser is required; and the mulch of crop stubble and dead weeds holds the soil together and retains moisture in dry weather. Perhaps one-third of US farmers are involved in some sort of conservation tillage, and the figures are rising. This is hardly 'organic', in that it relies heavily on herbicides to kill off weeds; but it is a sort of halfway house in that it reduces the amount and cost of fertilisers needed and is kinder to the soil.

In Europe, a growing number of observers of the farming world have been pointing out the sheer unsustainability of the European Community's Common Agricultural Policy. The CAP consumes about three-quarters of the spending of the Common Market, and storing or destroying the surpluses it produces costs the Common Market £150 million (\$225 million) per week. Critics say that it makes rich farmers richer while leaving poor farmers poor; that its subsidies disrupt international trade and make it hard for Southern agricultural countries to export their own harvests for a reasonable price; and that the subsidies encourage the ploughing up of marginal land and the pollution of the countryside with nitrogen fertilisers, pesticides and herbicides – all to produce surpluses.

In 1985, the politicians seemed to begin to take this criticism on board: they put quotas on milk production – an odd place to start, according to many farmers, because dairy farmers cannot simply plough under a prize herd of cows developed over years and grow something else instead. But farmers saw this as a sign of things to come, and feared quotas on other produce in the future. (Supporters of organic farming point out that even if their approach is found to lower production by as much as 20 to 25 per cent, that figure very neatly matches the levels of EEC overproduction; organic farming could thus help to level the surplus mountains.)

How can the surpluses be reduced without destroying national farming systems and farmers? Howard Newby, the British sociologist and writer on rural life, predicted two possible paths which what is now known as 'farm adjustment' might follow in the near future. Agriculture in general could reverse the trend towards more and more intensive production and ever higher inputs. Harvests and production in general would be reduced, but less output would require less input in terms of investments, so farm profitability would not necessarily suffer. Newby does not say so, but this scenario offers great scope for organic methods. The other available option would be the concentration of agricultural production on ever fewer, ever larger, intensive farms, while marginal producers are allowed or encouraged to go out of business. Newby adds that the 'role of small farmers may have to be redefined'. Again, although he makes no mention of it, part of this redefinition could be a move towards organic farming. In 1986, British Agriculture Minister Michael Jopling announced the 'Setaside' scheme, whereby farmers in certain environmentally sensitive regions would be paid by the state to farm without using modern, high-input techniques. The scheme could have the effect both of reducing surpluses and of protecting scenic farming areas.

Of course, one key reason why a farmer will go organic is that he can usually sell quickly for good prices the vegetables, fruit or meat he produces. The organic farmers' suspicion that high-tech farming is unsustainable has been complemented by the growing suspicion in the minds of an increasing number of consumers that swallowing nitrates, pesticides and weed-killers is 'unsustainable' in terms of healthy living. In Northern countries there are enough consumers who can afford to back these convictions with cash and who are willing to pay extra in order to eat uncontaminated food. The extra cost involved for a family which decides to go organic in terms of its own diet remains hotly debated; estimates range from 10 to 50 per cent. If a family exists mainly on organic potatoes and bread, it will not pay too much over the odds; if its tastes tend more toward such delicacies as organic mangetouts and organic wines, then it will obviously pay a higher premium.

The higher prices of organic food seem to suggest that organic farming is and will remain a luxurious offshoot of mainstream agriculture, having little to do ultimately with feeding the world, especially with feeding the hungry poor of the South. But this need not necessarily be the case. Organic farming is becoming an international movement; US and mainland European organic farmers are more advanced than their British counterparts. The first two stated goals of the International Federation of Organic Agricultural Movements, which has eighty member countries both North and South, are: '1) to work as much as possible within a closed system and to draw upon local resources, and 2) to maintain the long-term fertility of soils.' (Lower down the list, with goal number three, the Federation stresses its desire 'to avoid all forms of pollution that may result from agricultural techniques'.)

Techniques which base agriculture on local resources and maintain the long-term fertility of soils are precisely the techniques desperately needed in Third World nations. In 1983, Tanzania's president Julius Nyerere opened a conference in his country on 'low-input farming' – farming using as little chemical fertiliser, pesticide and herbicide as possible. He invited experts, mostly from the North, to attend – not because he was concerned about Tanzanians eating polluted food, but because his nation simply cannot afford the requirements of high-input farming. He said in his opening address that when Tanzanian schools began to teach farmers about artificial fertilisers, they had stopped teaching them about composting, as it was 'old-fashioned'. 'The net result has been that in many places nothing is done to refertilise our soil after it has been used, much less to improve its fertility. Our peasants can no longer move on after their plot has lost its fertility; they just get less results from their sweat and – legitimately – complain that having told them to use fertiliser, we do not make it available at a price they can afford or when they need it.'

The problems of both the expense of pesticides and the resistance of pests are being dealt with more and more in the Third World by the use of 'integrated pest management', a 'semi-organic' control of pests by using their natural enemies and rotating crops, but also by applying pesticides less frequently and at optimum times. In China's Kwangtung Province ducks are used to control weeds, leafhoppers and planthoppers in rice paddies. Egypt has been releasing 140 million sterilised male Mediterranean fruit flies each day in order to keep down the numbers of these pests in vegetable gardens and fruit orchards, while similar experiments are being carried out with sterilised tsetse flies in Nigeria. Peruvian cotton farmers, by means of strategic timing, have reduced sprayings from twenty-six to four per season, while sprayings by Egyptian cotton growers have fallen from ten to four per season.

Thus researchers into tropical agriculture are turning, belatedly, to what might in the North be called 'organic', but in the South are called 'low-input' solutions: mixing crops to prevent pests from wiping out a whole field (something Africans did for centuries before colonial experts interfered); planting crops in alleys between lines of nitrogen-fixing trees; using green manure; and releasing natural predators of crop pests.

All of these methods go against the tide of 'Green Revolution' tactics, whereby new strains of wheat and rice are planted which use fertiliser and water more efficiently, but which thus require large amounts of fertiliser and water – usually in the form of irrigation – and which tend to be delicate and therefore need large doses of pesticides and herbicides. This approach has increased India's rice and wheat production dramatically, but the Green Revolution has been slow to spread. There are too many farmers in the world who can not afford to buy either the necessary inputs to grow the Green Revolution grain themselves, or the grain produced by these inputs. There is as yet no Green Revolution technology to apply to Africa's staple crops such as sorghum, millet and the rootcrops, namely cassava and yams. So while India, and the world, produces more food than ever before, there are more hungry people in India and the world than ever before.

The profits being generated by organic foodstuffs in the North will inevitably attract scientific research into a form of agriculture which science has traditionally found beneath its dignity. In time this could lead to new organic farming techniques, or perhaps even new crop strains, which will boost organic yields. And, if such research into low-input farming becomes fashionable in the North, this could provide a big psychological boost for such research on a worldwide basis.

'We certainly need more research,' said Eddie Fewings. 'But we shall get it as the organic movement spreads. I don't think anything like all farms will go organic. But I predict that, in time, 15 per cent of British farmers will be farming that way.'

By late summer of 1986, Eddie was not optimistic about his own future on Springhalls Farm. The hard winter and the non-existent spring had given way to a summer which seemed to end abruptly in mid-August with unseasonably cold weather, especially at night. Yields were down, and he was having to delay harvests of French beans, mangetouts, courgettes, potatoes and tomatoes. He was worried that if the harvests were put off much longer frost would catch the produce in the fields. He did not think that Springhalls' owner would be willing to come up with more capital to keep him afloat and to enable him to complete the transition to an all-organic operation.

However, he was surprisingly ebullient both about organic farming and his future in it. 'I have learned so much in one year. I have learned how to cope with the first flush of weeds you always get in the early spring. Given good weather, I can just cultivate them out of existence, encourage the crops to take over.'

If the worst came to the worst and he was forced to leave Springhalls Farm, Eddie had another farm lined up in northern England. He had also received a challenging offer to put his organic farming experience to work in the Middle East.

'Wherever I'm farming, I am going to be organic,' vowed Eddie.

Eddie Fewings amid his organic lettuce outside Reading. Ladybirds do the work of pesticides (above). *Colin Hutchins' Somerset farm: both organic and idyllic* (left).

Colin labours in his fields under the watchful eyes of a man from the Soil Association and a scarecrow (opposite)

Juliana Ross with news photographs documenting her difficult childhood (left). *Elaborate water-testing facilities at a Hewlett Packard plant, established after storage tank leaks across Silicon Valley in the early 1980s* (below). *A map showing the incidence of birth defects, stillbirths, illness and deaths in south San Jose 1981–2* (inset).

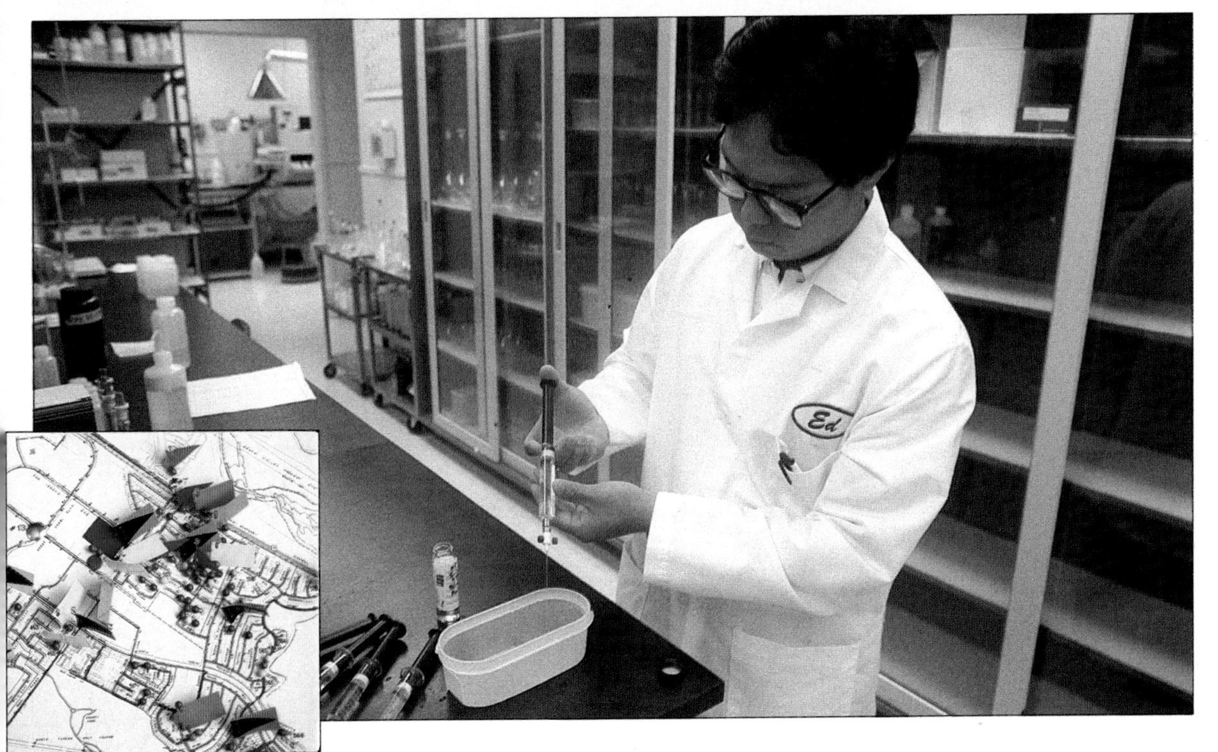

The landscape around Domboshawa village – good land over-farmed in Northern Zimbabwe (left). *Anna Ngwerume sweeps up amid the early morning shadows before fetching water* (below).

Anna and her family inside the cooking hut – the centre of family life (right). *Anna grinds maize on a stone as her cousin and sister-in-law keep her company* (below).

Anna supplements the education of her two younger sons (right). *Blazio and Jasmine, the younger of his two wives and mother of five, at work in the 'Struggling People' garden* (below).

A literacy class for women in Domboshawa, run by Anna, now an official 'Community Based Teacher' (left). *Domboshawa schoolchildren; Africa's population doubles every twenty-four years, faster than that of any other region* (below).

A woman's burden. Africa's women do much of the continent's work, raise most of its food, give birth to all of its children; they are the main hope for its future.

IX

CALIFORNIA: POOR WATER IN A RICH COUNTRY

'It seems depressing to me, but in the final analysis we are the only ones that can watch out for ourselves; there really isn't a lot out there protecting us from pollution.'
LORRAINE ROSS, MOTHER, Silicon Valley, California

Immediately after giving birth to her third daughter, Lorraine Ross checked the new baby all over, and counted her fingers and toes, to see if she were healthy and normal. Lorraine was relieved by what she found, as any mother would be. But this mother had more cause for concern than most, for two of her neighbours had recently had miscarriages, four others had given birth to babies with defects and one friend had had a stillbirth. These events had taken place over the year or so before Juliana was born in April 1981. It seemed a disproportionate number of childbirth tragedies among a group of women all living within morning coffee-visit walks of one another.

Lorraine Ross

Lorraine's relief was short-lived. Juliana slept too much for a healthy baby; she did not eat well, and she began to pant and perspire often. After visiting several local doctors in their town of San Jose, south of San Francisco Bay, the Rosses took one-month-old Juliana to a heart specialist. Lorraine vividly recalled that it was summer; she and her husband Jeff were dressed casually in shorts and had only a five-dollar bill between them. After examining the child, the cardiologist told the parents to take her immediately to the University of California Medical Center in San Francisco, an hour's drive north, without wasting time waiting for an ambulance. There the specialists diagnosed multiple heart defects and operated upon Juliana until three o'clock in the morning. Jeff had to telephone his mother to ask her to bring him the money to get the car out of the car-park so that he could drive home. Lorraine spent the next four weeks with Juliana in the hospital. Juliana has since spent much time in doctors' surgeries and had open-heart surgery when she was three; she faces more such operations in the course of her life.

San Jose is one of about fifteen low-rise towns merging into one another in what used to be called Santa Clara Valley; they are not so much towns, but suburbs without a city. The valley was once one of the world's largest orchards; cherries, plums, apricots and walnuts were grown there for sale throughout the USA and abroad. When Lorraine was born in the valley in the early 1950s, it held only 300,000 people; when Juliana was born in the early 1980s, it held 1·35 million people, almost 8000 per square mile. Lorraine grew up in the valley, married her childhood boyfriend, Jeff, and they settled in the Los Paseos area of South San Jose. 'It was very fairytale-like, typical middle-class California. It has the reputation of being one of the planned communities in San Jose, and San Jose was an area known for its sprawl and not its planning,' remembered Lorraine.

Amid that sprawl is an industry which is the envy of the world. Santa Clara Valley has been unofficially renamed 'Silicon Valley', the centre of the US and perhaps the world's microchip and computer industry. There are actually more such high-technology jobs in Los Angeles, to the south, but Silicon Valley has the highest concentration of these jobs and these companies – perhaps as many as 3000 firms, according to one estimate. These businesses do not only make microchips and computers; they produce telecommunications systems, medical hardware, laboratory equipment and office mechanisation devices. Before a slump affected the computer industry in 1985, there were four or five new companies setting up in the valley each week. No one was counting how many were moving out. There are over 120 factories producing microchips for computers. Most of the factories of this industry are clean, low-rise complexes made from glass, marble and steel. In fact, they look more like modern university science buildings than manufacturing plants. The plants are not even called 'factories'; they are known as 'facilities'. There are no smoking chimneys, and the manufacture of tiny silicon chips does not produce vast quantities of unsightly waste matter. Inside, many of the workers wear uniforms which make them look like surgeons; they work in almost dust-free 'clean rooms' designed to keep foreign bodies out of the microchips' intricate circuits. At lunchtime, they eat their sandwiches on the landscaped and well-watered lawns of their facilities.

It appeared to Lorraine, Jeff and their neighbours that the perfect 'sunrise industry' had come to this sunbelt area. It fitted in with the jogging, swimming, cycling life-styles of California; indeed it made those life-styles affordable for many. Household incomes in Silicon Valley are among the highest in the USA

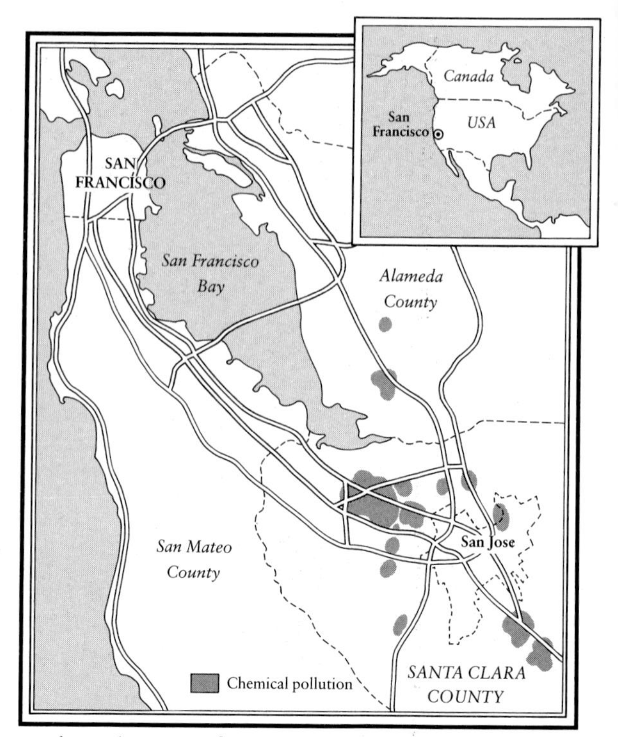

at about $30,000 (£20,000) per year. Not everyone is employed in the microchip business but the industry produces 200,000 jobs for the 1·35 million residents. Jeff Ross teaches in a local school, is a swimming coach, and drives a lorry part-time to make ends meet, but the Rosses claimed their share of the fairytale, and joined the local Cabana Club to swim and play tennis. They were not very upset when the Fairchild Camera and Instrument Corporation opened a 200,000-square-foot plant near their house in 1975. Lorraine remembered looking through the eucalyptus trees at the plant and wondering if they made good cameras. Actually, Fairchild, which later changed its name to Fairchild Semiconductor, was not making cameras. It was making silicon chips – turning sand into the brains of computers.

Then on 20 January 1982, the *San Jose Mercury News* daily newspaper, which has been covering events in northern California since it was founded in the gold-rush days of 1851, carried a report which said that industrial chemicals had been discovered in a well which supplied drinking water to 16,500 homes in South San Jose. The chemicals had been traced to

The Ross family at home in Silicon Valley, with Juliana on her mother's lap

a leaking underground storage tank at the Fairchild plant, 2000 feet from the well. Workers had found the leak by accident when they were replacing a tank and noticed discoloured soil. The tank's level indicator had been giving incorrect readings for some years, so it was not clear how long the tank had been leaking or how long it had taken the chemicals to infiltrate the well. The well had been closed down on 5 December, but the public had not been informed because, as one state health official said, the contamination was not considered to be a 'major threat to public health'.

The *Mercury News* had learned of the incident through an anonymous tip-off that 'Fairchild and the water company were poisoning the water in South San Jose,' recalled general reporter Susan Yoachum, who had been assigned to follow up the story. 'I thought, "Oh sure!" I thought it was a nut call; nut calls are an occupational hazard to journalists. But I called up the Great Oaks Water Company and found that Well Number Thirteen had been found contaminated with chemicals.' The water company had not wanted 'to announce anything until we knew what the problem was'. Susan's story appeared on the local news pages inside the newspaper.

Lorraine Ross read the article. In a way, she had been waiting for something like this to happen: 'We mothers would often get together and wonder why our children had birth defects or why we had lost babies. We would banter around theories that maybe Los Paseos had been built on a toxic dump or something; we had nothing but our imaginations to go on. We didn't know why we all felt a little guilty that it had happened to us, or what we had done. I remember sitting at the kitchen table, looking through the headlines and small ads, and buried in the local section was this small article saying that 1,1,1-trichloroethane, a chemical I couldn't pronounce then, had been found in the water in the Great Oaks system. I said, "Wait a minute; that is where I live!"'

Lorraine rang Fairchild but could not get through because the lines were jammed with other callers. She rang Great Oaks Water Company president Betty Roeder, telling her about Juliana's birth defect and the tragedies which had struck her friends. Mrs

Roeder confirmed that water from Well Thirteen had been piped into the Rosses' house; she encouraged Lorraine to write down her worries and send copies to her, to the health authorities and to Fairchild. Lorraine also telephoned Susan Yoachum, who also requested a copy of the letter. Susan remembers that she was almost as sceptical of Lorraine's account of the birth defects as she had been of the original tip-off but, when she checked with the health authorities after they had received Lorraine's letter, she found that they were taking it seriously and that investigations were being carried out. Susan wrote a follow-up story, quoting Lorraine's letter. It appeared on the front page on 2 February, with the headline: 'ELECTRONICS PLANT, BIRTH DEFECTS LINK?'

Viewed from the brutal perspective of international statistics, the story hardly merited front-page coverage. Every year, an estimated 10 to 25 million people die from diseases caused by unclean water. Every year, approximately 6 million children under the age of five die from diarrhoea usually traceable to bad water – that is 16,000 children per day. Were a single disaster to kill that many children in a day, the world's media would carry the story for weeks. However, this widespread carnage rarely makes headlines, or even back-page fillers, because the dead are the offspring of poor people in the under-developed nations of Africa, Asia and Latin America.

However, this was the USA and it was one of the wealthiest districts in this wealthiest of nations. As Lorraine Ross herself said later, 'Until something really hits you, until your kid gets sick, most people don't pay attention.' In her letter to the water company she had referred to the health authorities' calming statements quoted in Susan's first article: 'This may not seem to be a "major threat to public health", but it is of major importance to us who are so directly affected. We are fearful that twenty years into our future, we may be experiencing repercussions from this.'

At 7.30 am on 2 February, Lorraine finished reading the story in the *Mercury News* of a possible link between birth defects and the local water supplies. Because the article mentioned Lorraine's case and the cases of her neighbours, 'My phone rang and it never stopped ringing for two solid weeks. Every time we put it down we would have someone calling us and telling us that they wanted to be put on "my list". I really didn't have a list besides what I put in the letter, but these callers lived right round the corner from me and they had had a child with a birth defect, or lost a baby or had had a miscarriage – just call after call after call. At the end of one week I had a list of seventy-two names.' The same thing happened at the *Mercury News*, recalled Susan. 'The phones went wild. I've never seen anything like it either before that story or since that story. Another reporter was assigned just to field phone calls. Almost overnight it became a national story.' A map was organised to identify South San Jose health problems: red triangles for a baby with a heart defect, blue pins for miscarriages, yellow pins for cancer, thirteen black flags for fatalities.

The national and international media descended on this area which Susan has dubbed 'as middle-America as it comes: cookie-cutter houses, nicely manicured lawns, station-wagons in the driveways. It is in many ways a picture of America as I think about it in the 1950s rather than the 1980s. On the other hand, the high-tech industry is the hope for the future not only of America but of the world. The idea that those two could collide in a way that poisoned the people's water was something beyond our wildest dreams. Nobody even thought it was possible. We approached the story with a sense of amazement that this was going on.' Susan was speaking not only for her paper but for the media in general. Newspaper journalists, radio reporters, television crews, the international news services all arrived on the Rosses' doorstep to interview Lorraine and to film and photograph Juliana. Jeff remembered: 'We'd never been on radio or TV and all of a sudden we had five major radio and television networks on our front lawn every single day for about three weeks. I could not even come home the normal way; I had to hop over the fence just to get away from them.' Lorraine grew so hoarse from speaking to journalists and concerned neighbours that she could hardly talk. Her only previous experience with the media had been ten years earlier when she had sent an announcement of her wedding to the *Mercury News*.

Lorraine Ross is a strongly built, energetic and articulate woman with shoulder-length blonde hair and a dominant chin which juts out when she is angry. She was angry and ready to fight. 'When the cardiologist had told me what was wrong with Juliana, I had cried for three days non-stop, and I think that helped me deal with everything else that came after, because I think I got all the sorrow out at that time. Overnight I was thrown into a situation where I would not have wanted to be. Thank God I am a fast learner,' she said. Susan Yoachum agreed that Lorraine had learned fast: 'In many ways Lorraine Ross is responsible for raising consciousness not only in this community but throughout California in a way that will have an effect throughout the country.'

In early March, Lorraine, then aged thirty, confronted the San Jose City Council with the question: 'Fairchild or my child? In my mind, it is almost as simple as that. It takes a lot of nerve for them to invade a pre-existing residential neighbourhood, pour dangerous chemicals into a leaking tank, poison the surrounding environment and hide the fact from the people affected by their negligence.' She berated the Council for thirty minutes, and then she and her neighbours demanded from the Council answers to sixty written questions. All the questions boiled down to asking 'How could this happen?' Her original question, 'Fairchild or my child?' began to appear on placards carried by mothers on demonstrations, and became the rallying cry of the campaign. Teenagers wore T-shirts bearing slogans such as 'Fairchild – death child', and the follow-up newspaper investigations into the industry appeared under headlines such as 'Clean Rooms, Dirty Secrets' and 'The Dirty Work on the Clean Chip'.

Electronics plants in Silicon Valley, once the nation's 'prune capital'

It was a long, frustrating campaign for the Rosses and their Los Paseos neighbours. The question of 'What really happened?' was almost as difficult to answer as the question 'How could this happen?' Fairchild admitted that it had discovered that the chemicals 1,1,1-trichloroethane (better known as TCA) and 1,1-dichloroethylene (DCE) had leaked from an underground storage tank into the groundwater and thence into Well Thirteen. TCA and DCE are two solvents which remove the grease from microchips after manufacture. The necessary environmental laws governing chemical storage and disposal had not kept pace with the rate at which the Silicon Valley industries were springing up because they were not thought to be polluting industries. Newspapers reported that the two contaminants in the well had been linked to effects on the human central nervous system and the cardiovascular system, but the amounts in the wells were in the 'parts per billion' range – that is, they were only droplets in huge amounts of well-water. The concentrations were higher than permitted by the US Environmental Protection Agency's guidelines, but the guidelines were for long-term exposure. Fairchild insisted that the leak had only affected the well for a few months – if that – before it had been closed down.

The water company had been required to test its wells only for bacteria, viruses, pesticides and herbicides, not for industrial chemicals. However, in the summer of 1980, the state health services had tested all the major public wells in the area for such chemicals; the sample bottle from one of those wells had been either lost or broken, and the well had not been retested. The well in question was Well Thirteen in South San Jose.

The most important fact to emerge was that these chemicals had never been scientifically connected to birth defects. A Fairchild spokesman said with some justice: 'We are very strongly hopeful that there is no cause-and-effect relationship. Based on all the medical information at our disposal, we feel that there is no evidence to support such a claim.'

Health officials began to carry out studies. Finally, in January 1985, the results of two separate studies were revealed. Yes, there had been more heart defect cases in South San Jose babies than in the rest of the county, but this was qualified by: 'This investigation has not identified any environmental factor responsible ... In particular, the time and the geographic pattern of cases makes the contamination of Well Thirteen by the Fairchild leak an unlikely explanation.' The other study admitted: 'Spontaneous abortions and congenital anomalies in the Los Paseos area did occur above the expected rate,' but the evidence 'was insufficient to determine whether the leak of chemicals into Well Thirteen caused the excess.' In other words, after three years of study the authorities were able to reveal only that something odd had happened in Los Paseos, there being no proof either that this had been caused by chemicals, or that the chemicals were not to blame.

'Three years to verify the fact that there were three times the normal amount of birth defects in this neighbourhood,' marvelled Lorraine Ross. 'Well, we already knew that.'

Again, this being the USA, lawyers were called in. Local residents jointly sued the water company, Fairchild and the companies which had constructed the leaking storage tank. The water company sued Fairchild. Fairchild sued ten companies involved in the design, planning, construction and manufacture of the tank. By early 1985, Fairchild faced more than 400 lawsuits. IBM, which was later found to have also contaminated the groundwater, was added to the residents' suit. Fairchild replaced the fibreglass tank which had leaked with a steel tank in a concrete vault. In 1983, the company closed down its relatively new South San Jose plant – not for environmental or health reasons, a spokesman insisted, but because it was developing a new plant in Washington State to the north. He went on to say that by 1986, Fairchild had spent more than \$25 million (£17 million) to clean up the site around the closed factory, and proposed spending a further \$5 million (£3·3 million) to sink clay walls thirty to forty feet down into the ground to contain any chemicals still in the soil. The company had also dug a series of test wells to trace groundwater movement, and was pumping out groundwater to channel the seepage of chemicals back toward the tank rather than away from the plant.

The publicity campaign and the slogan 'Fairchild or my child?' had tended to single out the one company. But after the Los Paseos occurrence, 'Other companies began to check their storage tanks and almost every one found a leak. The number of leaks was astonishing,' said Susan Yoachum, who had found herself permanently assigned to the story. 'It was not the sort of story where you had to go to the office and ring around and see if there was anything new going on. Every morning when I came in there were messages waiting for me about new developments.' Later the authorities tested seventy-nine companies, and found that sixty-five of them – 82 per cent – had dangerous chemicals in the ground beneath their plants. All of the major names in microchips and computers – IBM, Intel Corp., Hewlett Packard, Tandem Computers, Raytheon, NEC and many others – were involved. The official involved in the tests said, 'When you start getting nearly 90 per cent positive, it puts things into perspective.' At least four more public drinking water wells were closed down, as well as almost forty private wells.

According to Ted Smith, the head of a collection of environmental, labour and community groups calling themselves the Silicon Valley Toxics Coalition, there was only one consolation for residents sitting on what he called a 'toxic time-bomb': 'Fortunately, the deep underground aquifer that stores and supplies water for the bulk of the valley's residents still appears clean.' That was in 1985; in 1986, county officials announced that 'relatively high concentrations' of chemicals thought to cause cancer had been found for the first time in those deep aquifers.

Experts had thought that a layer of clay running under the county would keep the chemicals out of this deep water. However, the chemicals apparently entered the aquifer through abandoned farm-wells left over from the days when Santa Clara Valley was the nation's 'prune capital'. There were thought to be 10,000 such wells, but only 3500 could be found on maps, and many of these were beneath roads, factories and office buildings. It was impossible to fill them in. 'This is the one thing I have been afraid of the most,' said one county water expert.

At the same time as the residents were protesting, the microchip companies were being attacked from another quarter: their own workers. In 1980, a report made by state health officials had shocked Silicon Valley by revealing that occupational illness rates in the semiconductor firms were three times higher than in the manufacturing industry in general. Half of these illnesses were said to be connected with exposure to toxic chemicals. In 1981, the semiconductor industry had reported a 70 per cent drop in occupational illnesses, a remarkable achievement in the course of one year. In fact, the claim was so remarkable that workers, unions and journalists charged that the industry report was a cover-up. An investigation by a local television station found that the companies were not reporting all incidences of illness in their workers.

The electronics industry not only uses the degreasing agents which found their way into Well Thirteen, but hydrofluoric and hydrochloric acid to etch the microchips, as well as poisonous gases such as arsine and phosphine to give the microchips electrical properties. Arsine has the effect of rapidly destroying red blood cells in humans, and the only way to save someone exposed to a large dose of the gas is by giving the victim an immediate blood transfusion. The 'clean rooms' are clean not for the sake of the workers but for the sake of the microchips, according to Tom Fisher, a former microchip worker who claims to have been disabled by inhaling solvents: 'The people working in there are breathing these solvents over and over and over again. True, they're not breathing any dust, but they're getting lots of solvents. The cleanliness is there for the convenience of the microchip; it doesn't care if it is seeing solvents, because it's being washed in them. But the human lungs care.'

Environmental and labour activists pointed out that the majority of the people working in the 'clean rooms' were women, for the most part poorly educated black, Mexican and Vietnamese women earning relatively low wages. Company executives were not exposed to the chemicals. Some workers sued their firms; others took the side of the local residents in calling for a clean-up of the industry. Anita Zimmerman, another former worker who said she suffered from on-the-job exposure to chemicals, told a local residents' meeting: 'Our health is worth more than anybody's profit margin, and our children are worth more than anyone's microchips.'

Community fire brigades were forced to organise quick-response 'Hit Teams': firemen equipped with fire-proof clothing and breathing masks were specially trained to rush into high-tech facilities in the event of a spillage of some of the 3500 chemicals used in the plants. One fire chief said: 'We never know what to expect, what pipes have broken, what kinds of substances have gone crazy and how they are going to combine with other substances.'

The workers' concerns explain the reasons behind the unique partnership embodied in the Silicon Valley Toxics Coalition: local residents and environmentalists on the one hand, and unions representing industry workers on the other, both demanding tighter health and safety regulations and practices in that industry. Often in the USA, local residents and unions have conflicting interests. None the less, as one local labour leader said: 'It's not as unlikely an alliance as people think. The same chemicals that are polluting the public's drinking water are affecting workers on the job. If you define your goals clearly enough, you can get a lot of people working together.'

Despite the discovery of the Silicon Valley groundwater pollution, spokesmen for computer companies feel that they have been unfairly treated. Larry Holbrook, Corporate Environment Director at Hewlett Packard, said that press coverage had blown the problems 'a little out of proportion' compared with similar problems around the USA. The Silicon Valley story 'has received a great deal of focus, part of which probably has to do with the image of a high-tech, clean industry which suddenly has a toxic problem associated with it'.

Holbrook has a valid point. In 1985, the staid US journal *National Geographic* carried out a survey of hazardous waste across the nation, and found that since 1950 6 billion tons of it had been disposed of on US soil. Two-thirds of all the government-regulated waste comes from chemical companies. The US Environmental Protection Agency (EPA) had designated or proposed 768 waste sites for its National Priorities List by 1984, and expected that number to grow to 2500. California had sixty of the sites, but the East Coast took over: thirty-five in Florida, forty-nine in Pennsylvania, fifty-eight in New York and ninety-five in the small industrial state of New Jersey. The New Jersey gambling centre of Atlantic City had to find fresh supplies of drinking water for its 40,000 residents and its hotels and casinos when chemicals from a waste site were found in the city water supply in 1984. In 1983, the US government had to buy the entire town of Times Beach in Missouri for $33 million (£22 million) and move out its 2200 residents after high levels of dioxin were found in the soil, the result of spraying with contaminated oil ten years earlier in an effort to keep dust levels down.

However, it is impossible for the government to monitor the situation. Organised crime has been moving into the waste disposal business in the USA on a large scale, illegally burning, dumping, recycling and transporting poisonous materials to help customers avoid the high costs of safer, legal disposal. 'Organised crime is besting disorganised government,' according to one criminologist who studied the syndrome.

'I doubt if we'll come fully to grips with hazardous waste until more of us are affected by groundwater pollution,' Joel Hirschorn, director of a study of such waste for Congress's Office of Technology Assessment, told *National Geographic*. 'That will be traumatic for millions of Americans. It's one thing to hear about a dump across town and quite another to be warned not to drink the water from your own tap. The crisis is almost inevitable, because even if we had full compliance with EPA regulations, they don't protect our water or our health.'

US environmentalist groups accuse the Reagan administration of being slow to act in allocating the funds designated for the five-year, $1·6 billion (£1·07 billion) government 'Super-fund' which is designed to make the waste sites safe. EPA administrator Anne Burford and other high-level agency officials were forced to resign in 1983 after public outcry over EPA's sluggishness in dealing with these sites, which may eventually cost as much as $23 billion (£15·3 billion) to clean up. When the often poorly-regulated legal toxic waste sites become full, they leak into the groundwater. A 1986 California State Legislature report found that all nine of the state's major hazardous waste sites were leaking; three had been closed, and none of the six still accepting waste met state criteria for natural geological barriers – layers of rock or clay – to keep chemicals out of the groundwater. The report suggested that these legal sites were on the verge of becoming prime candidates for the EPA's priority clean-up list. The waste disposal system 'is actually a system designed to turn solutions into problems', the report said.

Given the scale and national scope of the crisis, what can ordinary citizens do? In the USA they can do what Lorraine Ross and her neighbours did: take legal action.

Soon after they learned of the polluted well in early 1982, the Rosses also learned that there was a statute of limitations affecting cases such as theirs: if they were going to take legal action, they had only one hundred days to begin the procedure. In April, they approached a young lawyer named John Tyndall, a former helicopter pilot during the Vietnam War, and he agreed to take on their case.

'They had no real axe to grind,' said Tyndall. 'They had no desire to make a lot of money in litigation. They merely wanted to know what had happened to the water and whether this was the cause of the problems with the children. True, we were suing the company for money damages, which is really about all you can do in California; we couldn't force a government clean-up operation.' The Rosses themselves admitted to a more practical goal: 'We knew that somewhere down the road something might happen to us and that Juliana was going to need more operations and might not be able to afford them. We wanted her to be taken care of.'

Tyndall began gathering documents and seeking witnesses. Fairchild and associated companies started to line up expensive experts to fight the case. In the three years of preliminary investigations, Tyndall collected '500,000 different documents and thousands and thousands of pages of depositions from witnesses'. He had expected the preliminary work to cost his side half a million dollars (£333,000), 'but in the end it cost a good deal more'. At the beginning of the proceedings he was representing seventeen people; by the end he was representing 530. None of these people were wealthy or could afford such costs, but Tyndall had agreed to do the work for a percentage of any final settlement. He estimated that the eventual court case and appeals would take a further four to six years and cost several million dollars. There was not even any certainty that the residents would win. The Fairchild tank had leaked and the pollution had reached the well, but there was no proof that the pollution had caused any health problems – even though a study by state health authorities had found births in Los Paseos back to 'normal' more than a year after the well's closure.

The case never reached the court. The residents did not want to suffer the anguish and expense of a long court case, and the companies apparently feared years of damaging publicity if the hearings and appeals dragged on. John Tyndall was personally acquainted with the lawyers representing the companies. An out-of-court settlement was reached.

In July 1986, Tyndall announced a 'multimillion-dollar' settlement between the companies – Fairchild, IBM, the water company and several companies involved with the leaking tank – and the 530 residents. The agreement called for secrecy concerning the actual amounts of money paid to the residents, but Tyndall pronounced himself 'surprised and elated that we were able to arrive at the settlement we reached'. He said that the awards were much bigger than the average $15,000 (£10,000) each received by the residents of Love Canal in New York, hundreds of whom had been forced to move after leaking drums of toxic wastes were found buried near their homes. The fact that the case never reached court meant that Fairchild and the other companies did not have to admit any negligence. Fairchild refused to comment on the case, and has declined to be interviewed by the BBC about any lessons it may have learned from the incident. In relation to another, more recent leak from a Fairchild plant near a different well, a company spokesman said: 'We found it; we reported it; we have a solution in progress. We are acting responsibly.' This seemed to sum up the company's feelings about the Los Paseos incident as well.

'It is not going to give Juliana back her health,' said Lorraine of the settlement. 'We are not going to be living on the Riviera or anything like that. But it's a settlement to secure her future, as her condition is such that she might not ever be able to work or support herself. Now we have a weight lifted off our shoulders as far as providing for her goes.' The Rosses had another child after Juliana, and now have four daughters to raise – Jennifer, Jamie, Juliana and Jody. Juliana outwardly appears a normal, healthy child, swimming and cycling with the family and teasing her baby sister. But she tires more easily than her friends, and must often stop playing in order to rest.

Did all the publicity, the newspaper articles, the television coverage, the lawsuits and the final settlement actually have any effect on the way things work in Silicon Valley? Tyndall thinks so: 'As a result of this case you will not see any tanks like the one that was put in at Fairchild's put in again. We now have some very rigorous ordinances here in Silicon Valley and throughout the state. Also, I think the governments, corporations and people all realise that a group like this, by banding together and going through the legal system, can obtain very substantial redress for this kind of a case.' Santa Clara County passed a 'model ordinance' covering storage tanks, which has been copied by cities and counties throughout the state. California became the first state in the nation to pass laws regulating underground storage tanks, and the state has recently been setting up a birth defect monitoring system.

'Most significantly, a problem which was not even recognised before 1982 has been found to affect more than 30,000 storage tanks in the state,' said Susan Yoachum, who now covers state politics in the capital at Sacramento. Covering politics in California means that she still writes a lot of articles about industrial pollution. 'The contamination of drinking water by chemicals is now considered by many to be California's number one environmental problem. It is in the news every day. It is the major issue in the [1986] state governors' race.'

Larry Holbrook of Hewlett Packard agreed that the industry has learned a lesson. 'Chairmen of the boards, company presidents have really got involved, wanted to become acquainted with the issue and lent their support to it. The Silicon Valley companies have spent in excess of $100 million (£66·7 million) in investigations, clean-ups and revamping of storage facilities. That is a lot more money than has been spent by any US government Super-fund.' Another company spokesman commented that the industry

The 1986 California governors' race. Toxic waste was a major issue.

was working to clean up the environment partly because in order to survive it needs to attract the brightest minds in the nation, and they are unlikely to move to an area where the water is not fit to drink.

Two factors encouraging the companies toward cleaner habits are the reputed large size of the Fairchild settlement, coupled with the fact that clean-up operations after the event are obviously more expensive than safe storage.

'The industry itself has begun to realise that the profit margins are going to be a lot greater if they invest a small amount into some safety precautions, not only for their workers but for the communities around them,' said Lorraine Ross. 'They don't have to dump their chemicals down unmonitored tanks. This is common sense for someone like me. But it seems to take some large board of directors' decision to make all this an important part of their future plans.'

It is hard to know whether anything has been gained in Silicon Valley. The industry says it has improved its practices. Many families live with personal tragedies which they still do not, and never will, fully comprehend. And the valley sits on a deep aquifer of water polluted by chemicals; the long-term and short-term effects of these chemicals on humans are not understood. Nor is it known when or exactly how these chemicals will reach the drinking water wells. 'I am not sure our water can ever be safe,' said Lorraine. 'We moved thirty miles from where Juliana was born, into the countryside, thinking we could enjoy a healthier life-style. Fortunately or unfortunately, the industry followed us down here. We are now drinking bottled water because our water has been contaminated.'

It is even harder to see what lessons the Silicon Valley story offers to the Southern nations only now developing their industries. John Tyndall, who has set up his own environmental law firm, is representing some of the victims of the 1984 toxic gas leak from a Union Carbide pesticides plant at Bhopal in India, an industrial disaster which killed 2500 people and injured 150,000. India is the world's most populous democracy and is a relatively open society, so when the US-owned Union Carbide leak caused deaths and unjuries, US lawyers were able to take up the cases of the victims in an attempt to win compensation. Most developing countries not only lack enforced laws on industrial pollution, but they lack a free press and legal system which gives victims the possibility of redress through the courts. They also lack the large numbers of non-governmental environmental watchdog groups, such as the Silicon Valley Toxics Coalition, Citizens for a Better Environment, Friends of the Earth and the Sierra Club, all of which were active in the Silicon Valley controversies. Furthermore, the victims of Third World pollution are often the very poor, the illiterate, those who are unable to take a stand against industry and government. What compensation has been awarded to those who sustained injuries when in 1984 petrol leaking from a pipeline exploded in Cubatão in Brazil, killing 500 people? Or to the 1000 families whose homes were destroyed when liquefied gas storage tanks exploded in suburban Mexico City in the same year, killing 452 people?

These, however, are examples of 'spectacular' pollution. Most industrial pollution, as in the case of Well Thirteen, is gradual and hidden. US environmentalist David Weir studied pesticide manufacturing plants throughout the Third World. He found in Indonesia a 'slow-motion' Bhopal: smoke from burning waste at a DDT plant had killed twenty-five local people over a nine-month period. He found in Brazil a chemical complex owned by a multinational firm which local people said had turned the nearby river red. They said the only 'emergency procedure' they knew of was to look at windsocks hung around the complex to find out which way to run. Several hundred metallurgical manufacturing plants have been set up on the shores of Sepetiba Bay, south-west of Rio de Janeiro. Shellfish from the bay, popular in Rio restaurants and a staple food of the local fishermen, have been found to contain ten to twenty times the legal Brazilian limits for zinc; fish and shellfish have been found with three to fifteen times the limits of chromium. Many unclean industries are moving to the Third World precisely because pollution regulations are lax. When the Kawasaki Steel Company moved one of its plants from Chiba in Japan to Mindanao in the Philippines, a company executive said by way of explanation: 'People in the Philippines don't know anything about pollution.'

Mexico City, 1984: liquefied gas storage tanks explode, killing hundreds of people and leaving 1000 families homeless – a spectacular example of Third World 'pollution'

However, they and other Third World residents are beginning to learn, with the help of such non-governmental organisations as the Centre for Science and Environment in Delhi, which has published two scathing 'Citizen's Reports' on 'The State of India's Environment', based on data gathered mostly from universities and the government's own, not very well publicised, reports. *Sahabat Alam Malaysia* (Friends of the Earth, Malaysia) and the Consumers' Association of Penang, both operating from the island of Penang off Malaysia's west coast, have published similar reports and constantly chide the government over its lax pollution standards. Both these organisations, and others now springing up through the South, know how to publicise their stories in the media and how to demand the attention of officials. But in too many Southern nations, persuading officials to listen remains far too dangerous an undertaking to be attempted merely to stop pollution.

Around the world, the frontline fighters against industrial pollution remain those most directly affected by it. These people – whether they be Indonesian villagers, Brazilian fishermen or the Rosses in California – are rarely the rich and powerful. The company chairmen and the government officials, on the other hand, own homes in safe, clean neighbourhoods.

'I just wish we could reach more areas,' said Lorraine. 'I wish that people all over the world would realise that if they want the industry, they are going to have to regulate it and keep watch on it. It seems depressing to me, but in the final analysis we are the only ones that can watch out for ourselves; there really isn't a lot out there protecting us from pollution. We have to be consumer watchdogs and to hold our governments and our industries accountable for what they do. Here was a small group of people in a small neighbourhood who screamed loud enough about a problem and got angry enough about a problem to get together and force some change. I don't think there would be any regulation of underground storage tanks in California today if people hadn't got together and said this shouldn't have happened and it should never happen again.'

X

ZIMBABWE: MANY CHILDREN, LITTLE LAND

'I have got it now! I am now able to help myself. No matter what happens, I can stand on my own two feet, knowing that I can do something for myself.'
ANNA NGWERUME, Zimbabwean mother of seven

Anna Ngwerume was a leader during the liberation struggle in Zimbabwe, providing food, shelter and information for the guerrilla fighters. She came through the war unscathed. It is motherhood which has almost finished her off.

Anna rises at five every day, before the sun; she sweeps the family sleeping hut, built of mud bricks and thatch, and then cleans the larger cooking hut, the centre of family life. She builds a fire, walks half a mile to the well to fetch water for tea and porridge, cooks breakfast for her family, and then often sets off for several hours of work in the fields. Anna is a tough, graceful woman. Her back is straight, her gestures flowing, almost regal. But she can also kneel before the cooking fire and remove a tin can of boiling water with her bare hands. She wears bright cotton dresses and manages to stay clean and crisp during a day of weeding, cleaning and cooking over an open fire. Her brilliant clothes stand out against the fields, bush, rock outcrops and huts, which are all various shades of brown. She invariably ties a crimson chiffon scarf over her short hair and low down over her forehead, which emphasises her smooth face and prominent cheekbones.

Anna Ngwerume

It is surprising that she has kept her good looks and even temper. She has three daughters, four sons and four acres of poor, dry land near the village of Domboshawa, forty-five miles north of the capital of Harare. Sometimes she also has a husband, Martin, but he went off to work in Harare for sixteen years while his family was growing and the guerrilla war was raging in the countryside.

'From 1964 to 1980, I had no husband. I was not getting any help with money,' said Anna. 'My husband would have a job; then he would lose it and come home. As soon as he got another job he would disappear again. Whenever he returned home, he would give me a baby and then leave. We would argue; I would ask him, "What can you be really thinking of, to neglect your wife and family like this?"' In Zimbabwe, as in much of Africa, husbands rule the family and when they are away their mothers take over, always

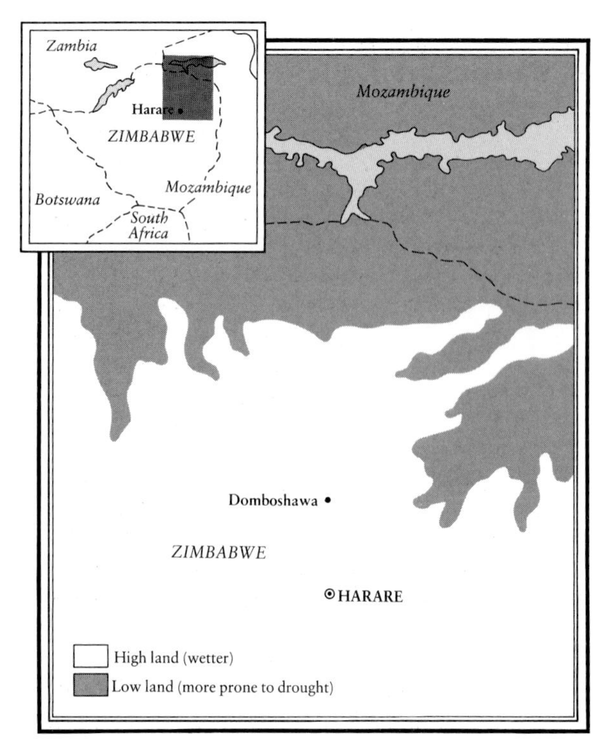

encouraging the wife to provide their sons and their clan with more children.

'My mother-in-law often said to me, "If you get this child off the breast, then I can take care of it and the other little ones while you have more." But then my mother-in-law died and, instead of her taking care of my children, I had to take care of all of Martin's eight younger brothers and sisters as well as my own children!' said Anna. 'I supported them all by raising maize to eat and green vegetables to sell.' Her last child, a boy, was named Omega – 'the end'. (Zimbabwean names often have a symbolic meaning; Anna's sixth child was born during the war and his name is Peacemaker.)

Many children, limited land and women trying to make ends meet. The issues of family planning, land use and women's rights are inseparable in Zimbabwe, as they are throughout Africa and the Third World. But Zimbabwe's efforts to get the balance right have been complicated by the fact that it gained true independence two decades after most of the rest of Africa.

Anna was providing for fifteen children at the height of the guerrilla fighting in the 1970s. At the same time, she was district vice-chairman of the Zimbabwean revolutionary forces. 'But then the chairman was forced to hide in the bush; so when the comrades [freedom fighters] came to the village, my house was their first port of call. They would ask me for food, money and shelter and I was expected to supply it.' She also provided information. The village children, Anna's eldest daughter Tracy among them, acted as the *mujibha* (eyes and ears) of the freedom fighters. During the day, when the white Rhodesian soldiers were in the village, the children behaved normally, playing games and going to school. At night they took food and news to revolutionary soldiers camped near by and guided them along village paths. Often the children were caught in the crossfire when the forces surprised one another; many died.

Zimbabwe's Independence struggle was largely about land. The country is a mixture of very good and very bad farmland. Eighty per cent of the land is over 2000 feet above sea-level, and thus has a healthy subtropical climate. But less than 17 per cent has good enough soils and sufficient rain to support intensive farming. Over 60 per cent is suitable only for livestock rearing; the rest can be reliably cropped only with help from irrigation and fertilisers. Much of the best land is on the high plateau around Harare, in the region where Anna lives, and which has a climate similar to that of southern California.

When Cecil Rhodes and his British South Africa Company marched into what was to become the colony of Southern Rhodesia in the late 1880s, they found more land than they could farm themselves. To encourage wage labour, they introduced a hut tax in 1893, and gradually began to expropriate the good land and to squeeze Africans on to poor 'native reserves'.

Soon the Africans had little choice but to work for the whites. The big commercial farms of the whites had tractors, high-yield crop varieties, pesticides and fertilisers. The Africans were crowded on to land which rapidly became overgrazed, deforested and eroded as they tried to make a living from it. The whites pointed to the ruined land of the Africans as proof that they were wasteful and ignorant farmers, even though many of the men, 40 to 60 per cent in some areas, were like Martin Ngwerume away from their own land working in cities, mines and on white farms.

The brittle bush and parched grass surrounding Anna's huts has a barren look, especially in the dry season before the rains of November to March. Over this harsh ground loom strangely shaped towers and cliffs of bare granite. Anna can remember when it was otherwise. 'There were many forests, huge trees and very tall grass with many wild animals: wild pigs, rabbits and small animals. In my young days the rivers were much deeper, with big pools we could swim in. Now the rivers are silted up; they are shallow and there are no pools. This is because many people farm along the rivers and cut the trees so the soil flows in. Today, there are few trees and we must walk miles into the hills for firewood. If the land had been this bad when I was a young woman, I would not have had nearly as many children.'

The 1980 agreement which gave Zimbabwe independence from Britain under a freely elected government stipulated that land must be bought, not seized, from white farmers. This was a blow to people such as Anna who had risked their lives for freedom and for land. But it was a pragmatic decision by Zimbabwe's new prime minister, Robert Mugabe. In 1980–1 the white farmers produced over 60 per cent of the national maize harvest, 72 per cent of the cotton and 90 per cent of the wheat. The whites argued that if they stopped farming, the nation would starve. So, five years later, there were still 4400 large commercial farms owned by whites; 2600 of these took up over three-quarters of the nation's best farmland.

Anna, along with well over half Zimbabwe's population of 8·5 million people, lives on former 'native reserve' land, now referred to as 'communal areas'. Almost three-quarters of this land is in regions where growing unirrigated crops is a very risky business. The nature of the risk became clear during the drought years of 1982–4, when 2·5 million people were forced to rely on government relief food for survival.

Anna's husband Martin began spending more time in Domboshawa in 1980. In 1983, Anna and Martin

A woman hoeing in the 'Struggling People' garden

helped to organise a co-operative gardening group of six women and five men. They intended to grow maize, tomatoes, onions, broccoli and lettuce during the dry season, hauling buckets of water from the nearby shallow stream. But their own investment of £4.20 ($6.30) each was not enough to buy the plough and all of the seeds and pesticides they needed. In 1984 they appealed to the Zimbabwe Women's Bureau, a government organisation which co-ordinates the work of women's groups, and they received a standard package of seeds and pesticides. The women worked daily in the 'group garden', which they named *Kutambura* ('Struggling People'). They sang as they hoed and weeded, but the drought persisted and the crops withered, leaving them with only a small harvest of summer maize.

Across Zimbabwe, others were forming similar groups and obtaining help from the government. Most small farmers were more successful than Anna's group, and the result was a spectacular turnaround in the harvests by African farmers on communal land. In 1978–9, these African smallholders had delivered to the national Grain Marketing Board only 3·8 per cent of the maize the board received; by 1984–5, these farmers delivered almost half the maize the Board bought. In 1985, the nation had its biggest maize harvest ever, with a surplus of 800,000 tons. It donated 25,000 tons and £667,000 ($1 million) to Ethiopia, making Zimbabwe the first non-oil-producing African nation to be both a giver, as well as a recipient, of aid.

Robert Mugabe's government had helped the farmers achieve this miracle by doing something that few other African governments have done: investing in peasant agriculture. Africa, where three-quarters of the people live by farming, was the only region in the developing world to decrease investments per farmer between 1978 and 1982. Zimbabwe's new government adopted the opposite approach, improving farmer training and advice schemes, providing small farmers with fertiliser and high-yielding maize seeds, and allowing them to borrow the money for these using future harvests as collateral. The government further encouraged farmers by raising the price it paid for maize from £22.35 ($33.53) per ton in 1978–9 to £75.95 ($113.93) per ton in 1985–6. The Grain Marketing Board increased the number of its buying depots, moving many into the communal areas to make it easier for farmers to bring their grain to market. The harvests prove that typical African subsistence farmers on small plots can produce high yields, given the sort of government help and encouragement most Northern farmers receive.

The government has not been able to afford to buy up white farms rapidly. By 1986, only 36,000 black families had been transferred to purchased white farmland, although in 1982 the government had pledged to resettle 162,000 families on such land. Populations in the communal areas are growing too fast for resettlement schemes to keep up.

Zimbabwe's particular population problems, a result of colonial policies, offer a distorted reflection of the more general population problems of Africa as a whole. When the 1984–5 droughts and famines struck more than twenty African nations, Zimbabwe included, many Northerners took the newspaper photographs of famine camps full of hordes of starving people to be proof that the continent was grossly overpopulated. They were both right and wrong. Africa has only about forty-seven people per square mile, far fewer than Asia's 264. But much of Asia is wet, and its population pressure long ago forced people to farm intensively, obtaining high yields of rice from small, well-watered plots. Almost half of Africa is too dry for cropping which depends on rain, and 80 per cent of the continent's soils suffer from fertility problems.

Africa's traditional abundance of land, much of it poor land, and shortage of labour and machines have meant that Africans need large areas from which to harvest relatively small amounts of millet, sorghum and root crops; they also need large families to work these large plots. Until the past few decades, growing families could simply spread out on to new land.

'The fact that Africa has had so much land is a population problem in itself,' said Dr Esther Boohene, head of Zimbabwe's National Family Planning Council and twin sister of Robert Mugabe's wife, Sally. 'We do not take into account whether the land is arable. People think: "But we have land, why can't we have as many children as we want?"'

Recently, much of the new land coming under the hoe and plough has been marginal land: steep, dry or barren. Per capita production of food crops has been falling drastically across much of the continent until, even before the drought, one-quarter of all Africans relied on imported food. A UN study suggested that only twelve out of Africa's fifty-one nations were capable of supporting their populations by their own food production without greatly increasing the use of high-yielding seeds, fertiliser, pesticides and machinery.

The falling food harvests have not encouraged Africans to limit family size. The continent has the fastest growing population of any region in the world, doubling every twenty-four years (compared with Asia's thirty-eight years), and producing 1 million new mouths to feed every three weeks. In much of the rest of the Third World, mothers are having more babies than they want; there is an unmet need for birth control. But in Africa, surveys found that women want about as many children as they are having, a continent-wide average of 6·43 per woman. Paul Harrison summed up the problem neatly in his book *The Greening of Africa*: 'The fact is that Africans plan their families just as actively as everyone else – and they plan large ones.'

The Shona people of northern Zimbabwe, as do many African cultures, practise traditional 'family planning' techniques based on abstinence from sexual intercourse, on withdrawal and on long periods of breast-feeding, which tends to prevent mothers ovulating. The goal of such traditional practices is not so much to limit the number of children, as to protect the health of both mother and child by spacing births. The Shona women, who carry their children on their backs, have a rule of thumb: if the child is big enough for its feet to cross around the mother's stomach, it is time to have another baby.

The reasons for Africa's high fertility rates are complex and include religious and cultural beliefs ('God gives children; who are we to refuse His gifts?'); the need for labour on the farm; the need for children to take care of parents in the parents' old age; and the extremely low status and lack of education of women in Africa. Bearing children is seen by many as the highest honour to which African women can aspire. It is also an obligation, certainly among the Shona people, where children are a way by which wives pay back to the husband's family the *lobolo*, or bride price.

'In some African cultures, if a woman dies without having given birth her dead body is beaten up,' said Dr Boohene. 'In other cultures the body is buried in a certain way because the woman has disgraced herself. In others it is given a special ceremony so the woman will not be reincarnated infertile again.'

Dr Boohene found the status of women to be lower and society in general to be more patriarchal in Zimbabwe than in her native Ghana. Only in 1982 did women have the right to vote and to be 'independent' at eighteen years old. Before that time, a woman could not even renew her passport without the signature of her husband or father. Women's income remains more heavily taxed than men's if their husbands are earning a salary, and this encourages them not to work but to stay at home and have children.

'The women played an important part in the war,' said Dr Boohene. 'They fought shoulder to shoulder with the men, learning how to shoot and live in the bush eating all sorts of animals. So they say that if today the nation is free, we should be able to enjoy Independence as much as the men.' They do not enjoy it. 'African women are always fatigued. They are constantly cooking, washing, cleaning, caring for their husbands. At the same time, they are pregnant. And by the time they start to have their fifth child, they are at great risk of losing their baby or their own life. In Africa, the number one killer of fertile women is pregnancy and childbirth.'

The white regime in Zimbabwe had set up family planning services, largely for whites, in 1953. During the liberation struggle, the guerrillas saw the government's efforts to extend this service to blacks as an effort to limit the number of blacks. After Independence, the white leadership of the Family Planning Association was replaced by blacks, but family planning was not a high priority of the new government.

Then from the chaos of hundreds of thousands of displaced persons caused by the fighting, a national scandal emerged. Young women were having babies, unclaimed by fathers, far from their families and villages. Many of these babies were simply abandoned in fields, gutters and rubbish bins; newspapers reported as many as thirty dead infants found in a day. This caused an uproar in a culture in which children are sacred.

The height of the 'baby dumping' scandal roughly coincided with the arrival in Zimbabwe in 1982 of Dr Boohene. In 1983, busloads of women arrived in front of her offices to protest the infanticide. They came 'not to ask why we weren't stopping it, but to ask what they could do to help,' said Dr Boohene. 'It was the first time African women had the courage to get up and say how they felt. They gave us a mandate to spread family planning everywhere in the country. It was the first time the government realised what the women of Zimbabwe really wanted. It saw its responsibility and began a total commitment to family planning.'

However, Dr Boohene had to find an African-style family planning approach. Trying to talk people into simply having fewer children would have met with great resistance. She named the new organisation the Child Spacing and Fertility Association in an attempt to convince families that the Association was not trying to prevent them having children but to encourage them in their traditional efforts of child spacing and also to help infertile couples. Only more recently have leaders begun to see rapid population growth as a threat to national prosperity. In 1985, Robert Mugabe summed up his nation's plight: 'With our large and rapidly increasing populations, with the expanding production facilities required to sustain such populations and improve their standards of living – in short, with the heavy pressure which is now continuously exerted on the resources of the environment – we face dangers which were quite unknown in the past.' To reflect this new view, the Association's name was changed to the Zimbabwe National Family Planning Council.

The rapid results of the Council's work over the past few years have been as spectacular as the harvests of the peasant farmers. In 1982, only 50,000 women were using contraceptives; by 1985, that figure had climbed to 400,000. This means 39 per cent of all Zimbabwean women of child-bearing age are using some form of family-planning technique, and 27 per cent of these are using modern contraception methods such as pills (80 per cent), diaphragms, loops and condoms. It is the highest rate in Africa, and far higher than in nearby nations such as Botswana and Kenya, where the rates of modern contraceptive use are 17 per cent and 12 per cent respectively.

The Council uses radio broadcasts in local languages and visits to schools, farms, mines, factories and agricultural fairs to proclaim its message. Popular disc-jockeys in Harare harangue listeners between records to 'start thinking about your family's future'.

But the backbone of the Council's programme is the network of 520 carefully supervised 'Community Based Distributors' (CBDs) of contraceptives and advice, most of them travelling on bicycles to villages each month. Almost all the CBDs are women, at least twenty-five years old, married with children and using contraceptives themselves. The service and the contraceptives are free to all with monthly incomes less than £63 ($94), the national official minimum wage, which is also the wage paid to the CBDs. (The average black annual income in Zimbabwe in 1984 was the equivalent of £170 ($255), despite official minimum wage laws, so many people are eligible for free contraceptives.)

The experiences of Anna and her friends in Domboshawa bring together the issues of land scarcity, women's status and family planning. Anna had stretched her meagre income to send two sons, Lovemore and Thomas, through four years of secondary school; her daughter Olivia was still in secondary school. Their mother had hoped to give her sons a brighter future than the promise of one day receiving a small piece of an already small farm. Today, Lovemore and Thomas are unable to find jobs, proof to Anna that there are too many people in Zimbabwe. 'The country is not getting any larger but the population is. It is a danger to the country and the country may not be able to improve,' she said. In fact, 48 per cent of all Zimbabweans are under fifteen years old, one of the highest percentages in Africa, so competition for both jobs and land will be fierce for decades to come. Anna was worried about Olivia: 'My daughter is in school now. If she cannot find a job, this will encourage her to get married earlier.'

The Kaviya family are members of Anna's gardening group. The husband, Blazio Kaviya, is a poor man, dressing in Western shirts and shorts which are no more than rags. His first wife, Sofia, could give him only one child, so he took another wife, Jasmine, who gave him five boys between one and thirteen. Jasmine is thin and anaemic, her face pinched. Anna was worried about her but did not feel she should interfere in the relationship between Jasmine and Blazio. However, as the fortunes of the group's garden began to look increasingly uncertain, Blazio began to gaze out more and more desperately across the dry vegetables and maize, and to compute income from the garden against the school fees he would have to raise as his children reached secondary school age. One day he came to Anna and asked her to help Jasmine obtain contraceptives. It was a village breakthrough: a husband had looked at his land, looked at his wife and family, and decided that contraceptives were at least part of the answer.

But it was a breakthrough of a negative nature: a man deciding that he simply could not afford more children for the moment. Dr Esther Boohene and her Council have more positive ideals in mind. Dr Boohene believes that if the status of women can be raised by teaching them to manage farms and money better, then they will feel themselves to be more than bearers of burdens and children.

Not only do African women give birth to all of Africa's children and do much of Africa's work, they produce, according to some estimates, as much as 85 per cent of its food. In Zimbabwe, the women grow this food on plots set aside for them from their husband's holdings. Even before Independence, the women had begun to set up farm and gardening clubs to use these plots to their full; they had also established other co-operatives to pool non-farming equipment, such as sewing machines, and also savings clubs to pool resources generated by the other work.

A survey of the Goromonzi district, in which the village of Domboshawa lies, found seventy-five such women's groups. The survey was the first step in a district-wide scheme, the Kubatsirana Project, to find these groups, improve the skills of the women running them, and see if they could be used to help in the spreading of family planning information and contraceptives. (*Kubatsirana* means 'working together' in the Shona language.) Its goal is to train 1300 women to act as Community Based Teachers (CBTs) to take their new skills back into their villages.

One day Elizabeth Muchatuta, field co-ordinator of the Kubatsirana Project, appeared in Domboshawa. Elizabeth explained the scheme to the local women's groups and asked them to elect a leader to be trained as a CBT. Remembering her wartime leadership, they selected Anna.

Elizabeth explained the reasons for teaching women leaders literacy training skills, management skills and family planning. 'Many of our women are illiterate but are courageously working in the interests of groups of other women to raise their standards of living. If they are educated, they can manage their projects better. They can cope with family planning better. They can use their education to convince their husbands that family planning is important. We can use these community based teachers to educate their groups and we can develop the whole district from just these small groups. Our main goal is to enable women to become self-reliant so they don't have to depend on others for their income.'

Anna's bus ride south of the communal lands to the CBT school was a revelation. One second she was in her familiar, harsh, brown landscape; the next second it was as if a wand had been waved and all the fields were lush and green, the houses big and modern and the farm roads leading from the highway paved and fenced. The bus had crossed the border between the old tribal lands and the big white farms. For decades, white farmers had been putting capital into large tracts of land: using irrigation, fertilisers, pesticides, contour ploughing and other anti-erosion techniques. For decades, the black farmers had been putting many children on to small tracts of land, with the resulting erosion, siltation and desertification. 'When I rode past those farms I wished I had something like that, so I could produce more of what those people are producing,' said Anna. She had learned a much sharper lesson about what capital and good management can achieve than any lesson she was to learn at the Kubatsirana Training Centre forty miles east of Harare.

Anna spent two weeks at Kubatsirana learning how to teach basic literacy, mathematics and book-keeping to others. She returned home for a few weeks to survey the needs of her own community and then spent another fortnight at the Centre learning to teach family planning. Another visit home was followed by a final period at the Centre to learn 'income generation techniques': development jargon for learning how to make and manage money. This training schedule was realistic, the organisers understanding that no village wife and mother could be expected to spend more than a few weeks at a stretch away from home.

When Anna and her twenty-six fellow students graduated, Sally Mugabe came to dance with the students in celebration. 'Comrade Sally', as she is known, urged the students to share their new knowledge unselfishly, as their teachers had done. 'Knowledge kept to oneself is not knowledge at all. I urge you to go back and use your knowledge to the benefit of everybody in your communities.'

When Anna, now earning a CBT salary of £71 ($106) per month, returned to her own community, her 'Struggling People' gardening group was at a particularly low ebb. The group garden had never done well because the group lacked the skills to organise effectively and to obtain sufficient loans. But they had recently made a little progress on their own. After unsuccessfully approaching several government groups for cash for irrigation, Anna's husband Martin had managed to obtain a loan of £6330 ($9495) from the volunteer agency OXFAM. The group bought a small diesel pump and some pipes and wire to fence the garden to keep out livestock. They dammed the stream themselves to form a small reservoir from which to pump water. However, they had planned poorly; other gardens upstream were also irrigating, so the pool of the 'Struggling People' was not getting enough water at the height of the dry season. The cabbage, carrots, tomatoes, onions, cauliflower, broccoli, okra and peppers were wilting. Anna and the group promptly drew up a budget of

Sally Mugabe (right), *wife of Prime Minister Robert Mugabe, and her twin sister Dr Esther Boohene*

£845 ($1267) for cement and other equipment to enlarge the dam to save the garden.

The change in Anna after her courses was much more immediate and obvious than any change in the garden. She had always been a strong, silent leader. Now she was an articulate, talkative leader with a purpose. 'I am anxious to set a good example by working hard so that the women can do the same and earn a good salary and be able to support themselves,' she said.

Where African women are usually painfully shy in discussing the intimate details of sex and family planning, Anna was suddenly frank and forthright. 'I have discovered that if I use family planning I can more easily satisfy my husband when he needs me without any fear of getting pregnant. The younger women using family planning feel more energetic because they are not giving birth frequently. It makes a great deal of difference to our appearances if we are not carrying babies on our backs all of the time; and our husbands find us more attractive, as we have time to look after ourselves. Not having babies all the time allows us time to take part in government programmes and to work harder for the groups. My own husband and I are now very close. We can sit and talk things over and make plans for the family. He even boasts about me with relatives and friends, even in the bars where he socialises.'

Changes in attitude do seem to be occurring, as women move from wanting only to space their children to wanting fewer babies. Anna was one of nine children, and her friends in the gardening group come from families of nine, ten or more. Most of Anna's generation have five to seven children. A recent

The 'Struggling People' gardening group builds a pump house to help with irrigation

national survey of fifteen- to nineteen-year-old women found that they wanted an average 4·6 children. Anna's daughter Olivia said that when she got married she wanted only four children. Olivia's sister Velma, aged fourteen, said she wanted only two. Such changes can have startling results in a few generations. If a woman has ten children, and if succeeding generations follow suit, she will have 1000 great-grandchildren. If she has four, she will have sixty-four great-grandchildren.

Zimbabwe must overcome other problems besides fertility to build a nation in which families such as the Ngwerumes can thrive. Even the huge maize harvests present challenges. Zimbabwe's agricultural policies have been praised by Northern aid organisations and UN agencies. Yet the policies of these same Northern governments threaten to upset Zimbabwe's 'maize miracle'. By the end of the 1986–7 marketing year, Zimbabwe was predicted to have stores of 1·7 million tons of maize, twice the necessary reserve against drought. In the southern African region around Zimbabwe are seven countries needing maize, Mozambique, drought-stricken Botswana and Zambia being in particularly dire straits.

Anna's fourteen-year-old daughter, Velma, who says *she wants only two children when she marries*

Despite these facts, Northern grain producers continued to send food aid to these nations, rather than sending cash enabling them to buy grain from neighbouring Zimbabwe. Equally surprising is the fact that if the hungry nations were to buy maize from the United States at market prices, they could import it more cheaply than maize imported via the poor rail-systems out of landlocked Zimbabwe. The US and the EEC undercut the efforts of efficient producers such as Zimbabwe by 'disposing of grain at subsidised prices', complained Zimbabwe Finance Minister, Bernard Chidzero. Other officials suggest that the US and other donors are more interested in dumping their own large surpluses than in funding projects which encourage the efficient use of surpluses in the region. Zimbabwe, which had been growing food efficiently in a hungry continent, may find itself exporting maize at great loss to its national treasury.

Back in Anna's village, her own daily round had changed since her training. She still rose before dawn, swept, cleaned, fetched water and cooked breakfast. But she was then more likely to talk to women in their huts and their fields about family planning than to work in the fields herself. She also taught reading, writing and book-keeping to women in local schoolrooms.

The fortunes of the Struggling People's garden remained precarious as the group continued to seek a favourable loan to enlarge the dam. Anna was not sure that they could do this in time to save the harvest. But she was sure that she and her friends would make the right choices eventually. 'I went to school when I was young and then I had a family. Only when I returned to school did I realise how much knowledge I did not have. I have got it now! I am now able to help myself. No matter what happens, I can stand on my own two feet, knowing that I can do something for myself.'

There are many enemies to sustainable development: drought, poor soils, pests, disease, warfare, superstition, short-sighted politicians and inequitable terms of trade. But two of the most persistent enemies are fatalism and resignation. Anna, mother of seven, is no longer fatalistic. She sees possibilities, recognises her ability to make choices and decisions, to raise questions and improve her life and the lives of others. She shares this new freedom with other people described in this book: the Sri Lankan monk, Kiranthidiye Pannasekera Thera; D'igir Turoga in Northern Kenya; Dudley Tapalia in the Solomon Islands; Emerita Castro in the shantytown of Villa el Salvador outside Lima; and Li Guangming in China. In the North, Eddie Fewings in England and Lorraine Ross in California also seized the initiative when faced with destructive forms of agriculture and industry.

These people have little in common; they live in different environments, operate by different cultural guidelines and face different challenges. The efforts of all may eventually be overwhelmed by forces and events beyond their influence. But all are working to make the most of the resources around them and to guarantee a future for the generations to come which will need those same resources.

'I have got it now! I am now able to help myself,' cried Anna, proudly. In helping herself, she is helping her family, her neighbours, her nation and her planet.

'I have got it now!'

BIBLIOGRAPHY AND SOURCES

BLAIKIE, P. *The political economy of soil erosion in developing countries* London: Longman, 1985.

BORAIKO, A. '*Storing up trouble ... hazardous waste*' in *National Geographic* March 1985. pp. 318–51.

BULL, D. *A growing problem: pesticides and the Third World poor* Oxford: OXFAM, 1982.

CASSEN, R. et al *Does aid work?* Oxford: Clarendon Press, 1986.

CHAMBERS, R. *Rural development: putting the last first* London: Longman, 1983.

CHAMBERS, R. 'Normal professionalism, new paradigms and development' A paper for the Seminar on Poverty, Development and Food: Towards the 21st Century, Brighton 13–14 December 1985.

CHAUHAN, S. K. et al *Who puts the water in the taps?: community participation in Third World drinking water, sanitation and health* (An Earthscan paperback) London: International Institute for Environment and Development, 1983.

ECKHOLM, E. P. *Down to earth: environment and human needs* (Foreword by Barbara Ward) London: Pluto Press; New York: W. W. Norton, 1982.

ERLICHMAN, J. *Gluttons for punishment* London: Penguin Books, 1986.

GALBRAITH, J. K. Address to 1985 meeting of the US National Research Council's Food and Nutrition Board. Washington, DC.

HARRISON, P. *The Greening of Africa* London: Paladin; New York: Penguin, 1987.

KIRKBY, R. J. R. *Urbanisation in China: town and country in a developing economy, 1949–2000 AD* Beckenham: Croom Helm, 1985; Columbia University Press, 1985.

MYERS, N. ed. *The Gaia atlas of planet management* New York: Doubleday, 1984; London: Pan Books, 1985 (Ramphal quote).

MYRDAL, J. 'Europe and the Third World: some personal remarks' Address to Red Flag/Class Struggle Conference on the Third World, Oslo 15–17 November 1985.

NEWBY, H. 'Agriculture and the future of rural communities' Paper presented to The Other Economic Summit, London, April 1985.

NORTH, R. *The real cost* London: Chatto and Windus, 1986.

PYE-SMITH, C. and NORTH, R. *Working the land: a new plan for a healthy agriculture* Aldershot: M. Temple Smith, 1984.

REPETTO, R. *World enough and time: successful strategies for resource management* New Haven and London: Yale University Press, 1986.

SEYMOUR, J. and GIRARDET, H. *Far from paradise: the story of man's impact on the environment* London: British Broadcasting Corporation, 1986.

TERRILL, R. 'Sichuan: where China changes course' in *National Geographic* September 1985. pp. 280–317.

THEROUX, P. 'Great rivers of the world: the Yangtze' in *The Observer Magazine* 3 January 1982. pp. 6–17.

TIMBERLAKE, L. *Africa in crisis: the causes, the cures of environmental bankruptcy* London: Earthscan, 1985; Philadelphia: New Society Publications, 1986.

WARD, B. and DUBOS, R. *Only one earth: the care and maintenance of a small planet* London: Penguin Books, 1972; New York: Norton, 1972, paperback, 1983.

WEIR, D. *The Bhopal syndrome: pesticide manufacturing and the Third World* San Francisco: Center for Investigative Reporting, 1986.

WIJKMAN, A. and TIMBERLAKE, L. *Natural disasters: Acts of God or acts of man?* London: Earthscan, 1984.

World resources 1986 New York: Basic Books for International Institute for Environment and Development and the World Resources Institute, 1986.

ACKNOWLEDGEMENTS

A book which covers the planet must rely on the intellects and energies of many people. It is really the work of the men and women who tell their stories in these chapters, and I hope its spirit reflects their own. The book also relies on the vision of the late Barbara Ward (1914–81), former president of the International Institute for Environment and Development (IIED), whose 1972 book *Only One Earth* (with René Dubos) firmly and forever joined the issues of environment and development. Norwegian Prime Minister Gro Harlem Brundtland, chairman of the World Commission on Environment and Development and contributor of this book's Foreword, helps to carry on Barbara Ward's work today. The *Only One Earth* television series and book originate from the work of Richard Keefe of North-South Productions, Richard Sandbrook at IIED, Jon Tinker of the Panos Institute, Roger Laughton and Peter Firstbrook of the BBC and many dedicated BBC producers and researchers. Special thanks must go to Charlie Pye-Smith, researcher for both the book and the television series, and to Dr Phil O'Keefe, who provided guidance and checking.

The series was partly funded, and the book and the series given both ideas and impetus, by organisations such as UNESCO and its Man and the Biosphere Programme, UNICEF, the Norwegian Ministry for Development Co-operation, the Netherlands Ministry for Foreign Affairs, the Swedish International Development Authority, the Nordic Red Cross Societies, Redd Barna (Save the Children, Norway) and the Better World Society of Washington, DC, which produced one of the programmes.

Many photographers contributed, but two award-winning photographers deserve special mention: Mark Edwards of London and Barbara Pyle of Atlanta, USA, to whom we owe a special debt of gratitude.

PICTURE CREDITS

Pages 16, 17 & 18 Mark Edwards/Panos Pictures; 20 Popperfoto; 22 Mark Edwards/Panos Pictures; 23 Iain Guest; 24–5 OSF Picture Library; 27 Aspect Picture Library/J. Alex Langley; 29 Iain Guest; 30 Picturepoint; 33 Aspect Picture Library/Tom Nebbia; 34 Trygve Bølstad; 35(t) Aspect Picture Library /Tom Nebbia; 35(b) Mark Edwards/Panos Pictures; 36 & 37 Aspect Picture Library/John Wright; 38(t), 38(b), 39 & 40(t) Iain Guest; 40(b) Tropix Photographic Library; 41 Anthony Geffen/BBC; 45 & 47 Mark Edwards Picture Library for Publishers; 49 Anthony Geffen/BBC; 50(t) & 50(b) Mark Edwards Picture Library for Publishers; 51(t) Anthony Geffen/BBC; 51(b) Mark Edwards Picture Library for Publishers; 52 Anthony Geffen/BBC; 53(t), 53(b), 54, 55(t), 55(b), 56(t) & 56(b) Caroline Webb; 57 & 59 Anthony Geffen/BBC; 61, 64, 66, 68 & 70 Caroline Webb; 73, 75, 76, 77 & 82 Anna Lloyd/BBC; 85 & 86 BBC; 89 Caroline Webb; 90, 91, 92(t) & 92(b) Anna Lloyd/BBC; 93(t) Nick Houghton; 93(b) BBC; 94(t) & 94(b) Dr Howard Reid/BBC; 95(t) & 95(b) Charmaine MacCormack/BBC; 96(t) & 96(b) Dr Howard Reid/BBC; 97 Nick Houghton; 100 & 102 BBC; 105 Charmaine MacCormack/BBC; 107, 108, 111 & 113 Dr Howard Reid/BBC; 117, 120, 121, 123, 125, 129(t), 129(b) & 130 Mark Edwards Picture Library for Publishers; 131(t) & 131(b) BBC; 131(inset) Charmaine MacCormack/BBC; 132(t), 132(b), 133(t), 133(b), 134(t), 134(b), 135(t), 135(b) & 136 Barbara Pyle; 137 Jonathan Darby/BBC; 139 BBC; 141 J. Allan Cash Photo Library; 146 Jonathan Darby/BBC; 148 Associated Press; 149, 151, 156, 157 & 158 Barbara Pyle.

Front jacket photography by Trygve Bølstad; back jacket photograph by Barbara Pyle.

INDEX

Page numbers in *italic* refer to the illustrations